THE
REASSEMBLER

JAMES MAY

HODDER &
STOUGHTON

'My first reassembly experience was the one that many of you will have suffered since, even in adulthood. That is, it didn't go back together.'

INTRODUCTION

I hope you haven't opened this book expecting an old fart to say *Back in the good old days things were made properly and designed to come apart. None of this 'no user serviceable parts' nonsense. Held together with proper self-tapping screws they were and so on and so on.*

Here's the truth of it. In those supposedly 'good old days' things had to come apart because they kept going wrong. Stuff was expensive, so had to be made to last, and that meant mending it. It's all very well moaning that people don't do their own car maintenance any more, but why should they? Cars work now, leaving you more time to spend with your family or learning to play the violin.

This copper and hide mallet is designed for knocking the wheel spinners off vintage cars. It's been used for pretty much everything else but.

So it doesn't bother me that my mobile phone doesn't come apart. By the time I bust it a better one will be available, which is how it should be. The total processing power of the whole world when I was born is now in my back pocket, ready to be sat on and broken, and it's come to that through the process of discarding the old and moving on. Taking things apart and mending them just holds us back.

However; this book is full of old things being reassembled because, well…

Old things were designed to come apart.

This is where I find myself impaled on the horns of a deep internal conflict. I love reassembly, and I'm a tool pervert. And yet, by rights, I should rebuild a 1950s lawnmower and then throw it in the canal, because otherwise I am reintroducing to the world the smoke-belching cacophonous evil that ruined a billion neighbours' sunny Sunday afternoons in the garden. A robotic electric hover mower is just better.

Fortunately, I have no desire to retard the world. Reassembly is merely a form of therapy; something that stimulates a part of my brain that is left wanting in my daily life. When I rebuild a bicycle, I re-order my head. So might you.

But that's all it is. I'm not an evangelist for it, I don't think it will save the nation, and it doesn't trouble me if the younger generation aren't interested – another common and pointless

A screwdriver with handy pocket clip, similar to the one used in alarmclockgate.

lament of the ageing. Why should they be? They've got Pokemon Go. I just happen to like putting things together. So I do.

It began for me around the age of five, and my first reassembly experience was the one that many of you will have suffered since, even in adulthood. That is; it didn't go back together.

The object of my enquiry was my parents' bedside alarm clock, and this being the late 60s, it was a proper mechanical device with a huge internal bell, and on which the time to get up could be set on a small subsidiary dial, to within an accuracy of 15 minutes.

The 'enabling technology' in this endeavour was the small, flat-bladed screwdriver (with handy pocket clip) that lived in a drawer in the kitchen (where else?). It was the first tool I understood. I pointed the business end toward the forbidden bedroom and advanced, undetected.

I can still see the three dome-headed and burnished slotted screws that secured the back of this clanking mystery. I knew, innately, that loosening them meant turning them anti-clockwise, even though I was as yet ironically unaware of what 'clockwise' actually signified.

Off came the back: suspiciously easily, I would now realise. More screws lay within, and on I went, like one of Carter's archaeologists, unable to resist the lure of the treasure that must soon be revealed to me despite a nagging sense that the whole project was cursed.

I did briefly glimpse the miracle of clockwork, with its gently oscillating balance wheel and train of gears, there to represent through simple arithmetic division the passage of our days. It had a beating heart; a tiny hairspring that contracted and swelled regularly to allow reciprocating mechanical action to assume the linearity of time's arrow. It lived! It was amazing.

But then it was gone. All of it. What I didn't appreciate, at the age of five, was the facility that the freshly wound mainspring had for completing the disassembly process for me, and as a quantum event at that. It was there, and then it simply wasn't there any more, scattered to the four skirting boards as the spring's impetus, intended to be metered out over 24 hours, was spent in a tiny fraction of time that a crude alarm clock could not in itself record. I was left astonished, clutching the empty shell of the ancestral timepiece, a cartoon boing hanging in the air over my father's approaching footsteps.

Even today, I approach a spring that is anything other than completely relaxed with deep circumspection, the legacy of a very early bollocking.

Meanwhile, an early distrust of electricity was engendered when I wrapped the wires leading to the working signal lights of my Triang trainset around the live and neutral pins of the plug on the transformer, and then pushed it into the socket. This

A bicycle chain rivet tool, similar to the one Cookie got for his birthday. Come to think of it, this may be the one Cookie got for his birthday. I might have nicked it.

electrocuted me. Still, the remainder of the trainset was a gloriously tangible thing held together with small brass screws, so would come apart safely.

There is, mercifully, only one spring in one of these old toy trains; the one that holds the brushes against the commutator of the motor. But then, the operation of the whole depends on it, and that, too, broke free with the determination of Airey Neave checking out of Colditz and was never seen again. I carried on anyway.

I suppose the remainder of the locomotive was unwittingly discarded by my mother in instalments, and one day she must have come across the empty plastic shell of the 4-6-2 Class 7P Princess Elizabeth and wondered where the rest of it was. Up the hoover was the answer, for while many people's childhood dreams were merely shattered, mine were victims of the abhorrent vacuum (cf the baddie figure from the Corgi James Bond Aston Martin with working ejection seat) because *I kept taking them to bits.*

It was with the bicycle that I finally achieved the satisfaction of what the Haynes Manual famously refers to as 'the reverse of the above', ie reassembly. It became clear to me and a handful of mates that a basic bicycle could be completely dismantled with just a handful of simple tools, up to 75 per cent of which we could muster between us. But not the rivet tool that allowed you to split the chain. Not even Cookie had one of those[1].

[1] He did get one some years later, for his birthday.
Life had little to offer in 70s northern Britain.

At first, just the odd wheel came off, or the brakes were swapped for a centre-pull set scrounged from a neighbour. But then we discovered the C spanner, so the bottom bracket could come apart. Then the cones and cages of a wheel spindle. Even the derailleur could be dismantled. And each time we dared ourselves to stray further into the mapless forest of disassembly, we found to our amazement that we could put it back together again and re-emerge, blinking, into the dazzling light of a complete bicycle.[2]

Soon, the naked frame hung from pieces of string from the rafters, all the other components bestrewn around it, except some of the ball bearings, which had rolled down the drain in the middle of the garage floor. Our bicycles enjoyed the use-to-maintenance ratio of a Formula One car. They were ridden furiously for a few hours, and then completely rebuilt, and for no reason other than because we could do it.

Bliss it was in those evenings for two or more of us to congregate like subversives producing *samizdat*; to strip a freewheel for sheer joy and then reassemble it packed with fresh grease so that it ran almost silently, even though it already did, because we'd done the same thing a week earlier. This was before we'd met any girls, obviously.

Real engineers and technicians would rail against all this unnecessary disassembly. It wastes time, it disturbs moving

[2] This metaphor needs taking apart and putting back together properly.

A scattering of bicycle tools. The funny spring thing is a 'third hand' brake adjusting aid. Fascinating.

parts that have 'bedded in', and it removes lubricant from places where it has established itself as a molecular component of the mechanism, and without which the whole thing might seize in a matter of minutes. But what could they know of the near orgasmic frisson we felt at being able to reverse a process that, until then, had seemed distinctly reductive?

It has to be said that, eventually, Cookie's bike was running with as few as three ball bearings in the entire steering head when there should have been around 50. We were dispatched by an irate father (his, this time) to buy a packet of replacements from the bike shop on the other side of town, but had to walk there, because our bikes were in bits.

Even today – old fart alert – I am outraged by bicycles that function in any way other than absolutely perfectly, because I know that with a few pressed-steel multi spanners made from cheese, a mallet, a punch and just one screwdriver that has at some time served as a chisel, they can be made to work with the unerring certainty of Babbage's Difference Engine.

The remainder of this rant about people not looking after bicycles properly has been deleted on the insistence of my publisher.

The digital revolution is a marvellous thing. The internet, I am convinced, is the most significant technical and social development of my life, by a huge margin.

But the world remains, in essence, a mechanical place, where 'machine' can mean something as humble as a tin-opener or as sophisticated as a perpetual calendar wristwatch. Electronic control has wrought wonders on the operation of, for example, car engines and even car suspension, but at heart these things are still packed with pistons, springs and levers.

Barring a few recent developments in three-dimensional printing, all mechanisms started life as components and required assembly. Dismantling them and reassembling them anew will reveal so much about how these things work, how they were designed and why they are the way they are.

You need to know these things, not least because if you understand the machine, you will be more sympathetic to it. Machines are like pets; if you're kind to them, they will indulge you more. The difference is that if your dog starts making a strange grinding noise after 20 years you can't strip it down and rebuild it.

Most importantly, though, reassembly is a benign form of creativity, because the outcome is pre-ordained. There is an interpretative skill to reading an exploded diagram, and there is bafflement along the way, but in the end it has either gone together correctly and works, or it has not and it doesn't. It's

This was discovered by me in the ruins of an abandoned factory years age. It has always been known as 'The useful block of metal'.

said that a great painting is never truly finished; a reassembly job is, and you will know that moment in the first pop of internal combustion or the first ascending glissando of a tightening guitar string.

I'm delighted that you are holding in your hands a book about putting things back together, because it's a subject that fascinates me but which I assumed was a lonely passion that I would take to the grave, unconsummated by the normal channels of human interaction.

Welcome! You and I, we are not alone, and our screwdrivers are our flashing Excaliburs as we sally forth to make small parts of the fragmented world whole again.

At the insistence of a bunch of old women I must point out that this book does not condone reassembly. Reassembly is a potentially dangerous activity. Failure to reassemble your food mixer may lead to marital breakdown, starvation, and death. The small springs in the Flying Scotsman could have someone's eye out. Reassembly is undertaken at your own risk.

'The trouble with the powered lawnmower is that it's actually a form of tyranny, and like most things in Britain its imagined appeal is rooted in the class system.'

1959
SUFFOLK COLT
PETROL LAWN
MOWER

(331 PARTS)

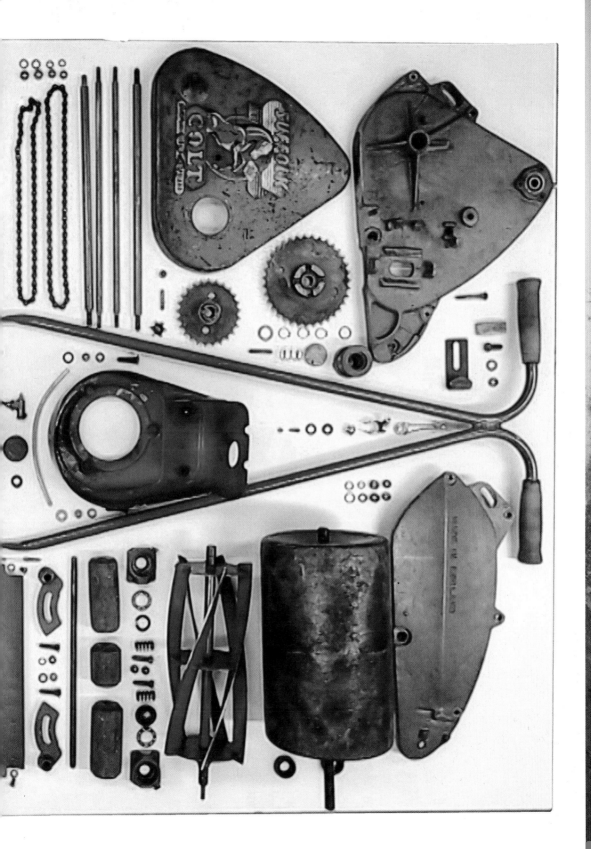

It has long been mooted that the public library will eventually disappear. Nobody borrows a book to read any more. But I see a new use for them.

How about a network of national reassembly vaults? You turn up with your library card, walk away with a box containing all 331 components of, say, a Suffolk Colt petrol lawnmower, put it together over the next two weeks and then take it back. The librarian dismantles it again and puts it back on the shelf, under 'L'.

For those without a shed or garage, it could be reassembled on the premises. It would be a great way to spend a cold and wet winter's afternoon, pretending to be interested in something while secretly saving on your domestic heating bill. Bit like a public library, really.

No need to source or actually own an old lawnmower, then. Serial reassembly could become an expensive addiction, and one that clutters your home with outdated artefacts that have no use and appeal only because they were designed to come apart

This is my poshest pair of pliers, and cost around £65 two decades ago. I had to stay in for a fortnight and just look at them, to offset the expense.

and are held together with proper
self-tapping screws [You did this in the
intro, ed].

The library lawnmower is already in
bits. If you bought your own Suffolk Colt,
it would almost certainly come intact,
and in dismantling it you would learn too
much and corrupt the delicious challenge
of trying to work out how it all goes
together, aided only by the single and
microscopic exploded diagram provided
by the sadists in the Suffolk Iron
Foundry's technical department.

Most importantly, though, the 'library'
system will prevent us repopulating the
world with revived Suffolk Colt lawnmowers.
A lot of people are deeply sentimental about
the Suffolk Colt. It undoubtedly enjoys some
significance in the history of post-war British
industry. There is also a curious sociological
angle to it, which we'll come on to. The
Suffolk Colt is a lovely thing to reassemble,
because it's not too fiddly and requires
very few tools. Educational, too, because it
introduces you to the wonders of the four-
stroke petrol engine in its simplest form.

But, as with any classic car, there's a reason it's not made any more. It's not because Honda and Kawasaki are evil, or because British society has gone to the dogs, or because Margaret Thatcher. It's because the Suffolk Colt is crap.

For a start, this great democratiser of domestic mowing has a blade width of just 12 inches. So even if you have only the modest front lawn of a typical British urban semi, that's a lot of upping and downing breathing cancer while your eardrums cave in.

Imagine, in the 1960s, when the Suffolk Iron Foundry was at the peak of its mower-making prowess, if every other home-owner in a 30-house cul-de-sac went out on Sunday afternoon to mow the lawn. The noise would be like World War One. We're not talking here about the delicious burble of a Venetian motor launch or the evocative exhaust note of a passing Merlin-engined Spitfire. We're talking about the moronic clatter of massed side-valve singles. Horrible.

These are external circlip pliers. They are not for use outdoors, but for removing external circlips. Can be used outdoors, though.

And for what? The trouble with the powered lawnmower is that it's actually a form of tyranny, and like most things in Britain its imagined appeal is rooted in the class system. It is like wanting a drinks globe or a lavatory inside the house, just much noisier.

Before the lawnmower, grass was cut with scythes by skilled men. Quietly. It was an expensive business that only the toffs could afford. The early lawnmowers were weighty contraptions designed to be pulled by teams of men or animals, and used for sports grounds, public parks and so on. It should be noted that a number of posh sports – tennis, bowls, croquet and cricket – required well-maintained grass, thus cementing the reputation of a neat lawn as a signifier of respectability.

Domestication of the lawnmower was a simple matter of miniaturising it. But then it was awarded the miracle of internal combustion. Now you, too, could have a closely cropped aristocratic lawn, and in the process of acquiring it appear modern and freed from labour. All you had to do was walk along behind the thing smoking a pipe; as well you might, since an industrial bronchial disorder was on the cards anyway.

As a result of all this, even slightly unkempt grass in front of a small family home came to indicate that the occupants were bad 'uns and would probably steal your bicycle. Long grass in a small English garden is as unacceptable as long hair on a guardsman.

Successive generations of innocent English families have been condemned because at some time the front lawn grew a bit too long. And that's my point. Having embraced the lawnmower, we immediately became enslaved by it. That the lawnmower existed as a simple route to social elevation meant you had to have one, and you had to mow the lawn, and thereby ruin the tranquillity for which the English garden, for centuries the envy of the sophisticated French, is supposedly famed. The English garden became a riot of din, fumes and mower-generated blasphemy.

I'm not suggesting for a moment that we

In the days before lying in adverts was banned, Suffolk could claim 'instant starting' even though the instant could last for up to an hour.

SUFFOLK 4-STROKES
INSTANT STARTING!

TRIED TESTED PROVED

...THE SUPREME 4-STROKE POWER MOWERS

THE 'PUNCH', 14" CUT
Entirely self-propelled ● Lower-than-ever maintenance cost ● Single lever finger-tip control ● Automatic clutch ● All transmission mechanism totally enclosed (safe and dirt-proof) ● 6-Blade Cutting Cylinder ● Self-aligning, twin race ball-bearings.

37½ GNS. incl. Grass Box (TAX-PAID)

THE 'SUPER-PUNCH', 17" CUT
A really "Super" 17 in. 4-stroke. Twin land-rollers with full differential mechanism. 6-Blade Cutting Cylinder.

46 GNS. incl. Grass Box (TAX-PAID)

THE 'SUPER-PUNCH' 17" PROFESSIONAL
For the Connoisseur and Green-keeper.

54 GNS. incl. Grass Box (TAX-PAID)

19" CUT

FROM ALL LEADING HARDWARE DEALERS AND AGRICULTURAL ENGINEERS
Over 200 Officially-appointed Service Agents throughout the country.

ALL 'PUNCH' MODELS NOW HAVE

★ **AUTOMATIC RECOIL STARTER**

★ **'DUAL-DRIVE'**
switching power from full drive to cutting cylinder only, at a touch.

★ **3-YEAR GUARANTEE**
for Engine AND machine—BOTH from "Suffolk" workshops.

THE 4-STROKE

'CORPORATION'— £29.18.6
with Recoil Starter

and **'SQUIRE'— £26.15.0**
Complete with Grass Catcher (TAX-FREE)
Centrimatic Clutch and safety cut-out device. For Orchard, Paddock, Verges.

FREE! Write to the Manufacturers for Illustrated Brochure giving full details of all Suffolk Mowers —Power and Hand:

SUFFOLK IRON FOUNDRY (1920) LTD
Winton House, St. Andrew Street, London, E.C.4

return to the days of scythe-wielding hardy peasants. I'm suggesting that maybe we cut grass a bit too much. Whence came the absurd notion that it should look like a National Service haircut? Nature doesn't do it, and long grass can be host to insects and butterflies the decline in which so many gardeners lament, as they mow the lawn. Again. Wild grass, after all, is like pubic hair, and seems to stop growing once it reaches a certain length.

I've always liked screwdrivers with translucent 'jelly' handles. They come in many colours and can be arranged attractively, like flowers.

And what, really, are we to make of a man who trims his minuscule lawn into contrasting stripes of uniform 12-inch width, as if preparing for an Elfin tennis tournament? That he is the sort of chap you can depend on in a tight spot? That the order apparent on his lawn is a reflection of the robustness of his debate? That he is neurotic, consumed by inner loathing, has never known real love and should never be granted a shotgun licence? I'm just beginning to wonder.

Tragically, I've also just remembered that this man was me. I popped my petroleum cherry at the age of 13 with, in case you hadn't guessed, the family lawnmower. It was an ancient Atco with the gratifying enhancement of twin bar-mounted clutches; one to engage the cutter, and then another to bring the roller into play.

The starting procedure for the Atco is so deeply embedded in the fabric of my being that I can feel the process in my fingertips as I think about it. Turn on the fuel with the tiny brass tap under the tank. Set the throttle open a touch. Tickle the carburettor float bowl (see panel) with the plunger until fresh and aromatic four-star seeped from the pin-hole in its side. Choke on. Roller clutch disengaged, the Atco equivalent of selecting neutral.

I've never found a specific use for these mini shears. Scissors or tin-snips always work better. But I felt sorry for them in the shop, so bought them anyway.

Grasp the burnished plastic[1] T-handle of the recoil starter with one hand, the nearest handlebar grip with the other. Then heave firmly and progressively for the complete length of the cord and that was it; the piston began its endless journey to nowhere up and down the bore of the engine.

[1] I'm not sure plastic can actually appear 'burnished', but you know what I mean. It was worn smooth and shiny by the sweating hands of men of toil long passed away.

What to do with this instant neighbourly dispute now it was running? Mow the lawn. It was useless for anything else, and attempts to repurpose it for towing a small bicycle or powering the chassis of an old pram came to nothing, and are forgotten footnotes in the history of personal transport. So I mowed the lawn. Endlessly.

On each day of those near-mythical mid-70s summers, the busy old fool[2] would rise over a lawn still reeling from the murderous passage of the mower the day before. I didn't know what sort of growth occurs in grass over one 24-hour period. Little if any, I now realise, but I wasn't having any of it. Up and down the antique Atco ran, the *dung dung dung* of its engine the sonic backdrop of all human awareness within a radius of half a mile, the grinning delinquent at its helm utterly unaware of the appeals to sudden and inexplicable death being heaped upon his straining shoulders.

[2] John Donne

This could be dismissed as yet another screwdriver, but it's never that simple. Width, length, handle shape and the type of tip lend infinite variety to the species. This one has a 'cabinet tip' if you're interested. Thought not.

Years later, I and the mower were separated by the normal course of home-leaving and study. Combustion was courted elsewhere, in cars and motorcycles, and the iron destroyer of peace fell mute and was eventually disposed of by my enemies in some garden tool barter deal. But recently, propelled by nothing more than sentimentality, I bought a restored 1979 twin-clutch Atco from the excellent British Lawnmower Museum in Southport.

Don't worry, though. I don't have a lawn.

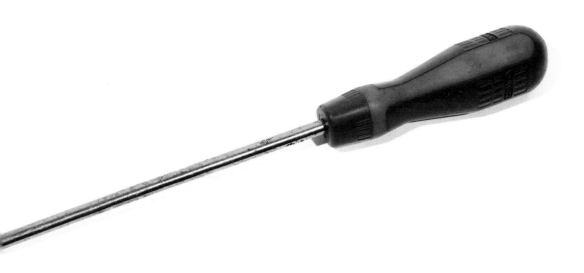

HOW THE CARBURETTOR WORKS

The purpose of the carburettor is to introduce fuel and air, mixed together in the correct proportions, to the cylinder of the engine. The word 'carburettor' means in essence 'adder of carbon to'; it's a device for adding fuel to atmospheric air. Engines run on hydrocarbon fuel of some sort but they need oxygen as well to achieve combustion. In a petrol engine the ratio is roughly 14:1 air to fuel.[1]

To understand the basic principles of the carburettor, you need look no further than the pioneering work of Giovanni Battista Venturi (1746–1822). His *Récherches experimentales sur le principe de la communication laterale du movement dans les fluids appliqué a l'explication de differens phenomènes hydrauliques* is a rollicking good read.

But in case you don't have time for all of it, the important bit to us here is the explanation of the *Venturi effect*. This says, in very simple terms, that if a fluid traveling down a

---

[1] This is why petrol is such a gloriously energy dense fuel for your car. You only carry about seven per cent of what you need in the fuel tank. The rest comes from the surrounding air.

My own diagram of the Venturi effect and how it applies to the carburettor. I wan't happy with any of the others I found. They must have been pretty bad.

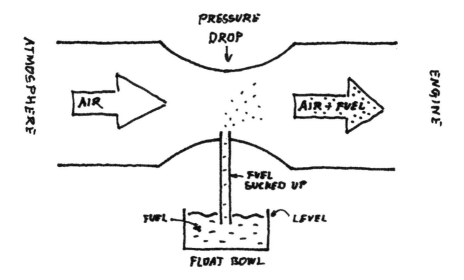

tube meets a restriction, it will speed up. The pressure at that point will drop, because the speed multiplied by the pressure remains constant.

Now then. We imagine that fuel for our lawnmower lives in the petrol tank and is fed to the engine from there. But that isn't quite true. There is a second, miniature fuel tank under (or to one side of) the carburettor, known as the float bowl. This contains a precisely metered amount of fuel maintained at a constant level. It's fed from the main tank via a cut-off valve not unlike the ball cock in a cistern.

The body of the carburettor is a Venturi; that is, a tube with a constriction in it. As the descending piston, on its intake stroke, sucks air through the Venturi a pressure drop occurs at the constriction. It follows that if a small tube is formed between the (low pressure) centre of the constriction and the float bowl, fuel will be sucked up.

It joins the torrent of incoming air to form an atomised spray of aromatic petroleum perfume, thence to rush unhindered into the cylinder, to be mercilessly sacrificed to the fire for the pursuit of our endeavours.

That's pretty much it, but the basic carburettor outlined above (which is how the first ones were) will only work properly at one engine speed. Engines need to operate over a range of speeds, so the carburettor requires several different fuel tubes (called *jets* in the lexicon of carburettorists), often including a variable one (a *needle jet*), and a slide to vary the severity of the constriction. By the time you arrive at a sports motorcycle engine, which might operate between 1000 and 15000 rpm, the carburettor becomes an instrument of incredible complexity and precision. There is usually one per cylinder, turning the whole 'stack' into a calibration nightmare or wet dream, depending on how good you are with very slim but very long screwdrivers.

The type of exploded diagram used in the reassembly of the Suffolk Colt. Reproduced here about four times actual size.

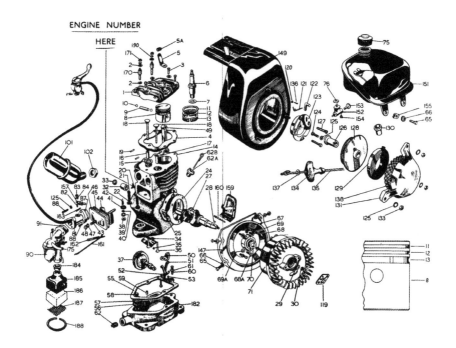

In recent years, the carburettor has been largely drop-kicked into the pedal bin of internal combustion history, to be replaced by more efficient and more minutely manageable fuel injection. Today, only basic garden equipment and simple mopeds use carburettors. Even Honda's most basic 125cc bike has fuel injection.

Bah. I like a good carburettor.

LAWN STRIPES EXPLAINED

It's not clear why stripes on a lawn are considered desirable but, as with most things lawn-related, there may be some snobbery involved.

Stylistically, the stripe has a chequered history. In the Middle Ages, according to the French social historian Michel Pastoureau, striped clothing was reserved for criminals, executioners, prostitutes, madmen and jesters. Stripes were a ready reckoner of undesirability, so no-one would have wanted them on the lawn, even if the mower had been invented.

Then again, stripes also came to be an essential ingredient of heraldry and military standards, all of which indicate high-class associations. And then, in 1846, the youthful Prince Albert appeared in a sailor suit[1], which is a bit stripey, and the sycophantic middle classes did the same with their own urchins, meaning stripes became a badge of good breeding. Perhaps.

[1] It was probably his mum's doing. That was Queen Victoria.

Actually, I think this is all nonsense, and that the story of stripes in fashion has no bearing on mowing. I think the desire for a stripey lawn comes from the home-owner wanting to boast that he has a *proper lawnmower*. That is, the sort of mower a pukka park keeper or groundsman would use.

A rotary hover mower, with its flat spinning blade, merely hacks grass down indiscriminately, like some grim reaper of the garden. But a cylinder mower actually *cuts* each blade of grass, from one edge to the other, as a pair of scissors would.

Note how the blades on the cylinder describe a gentle helix, and how they act against the fixed lower blade (part no F016101509), exactly as the two arms of the scissors would. This (apparently) gives the grass a neater appearance, for although you can't see the effect on the individual blade without a magnifying glass, it's perceivable from the overall impression.

Added to this is that the cylinder mower has a roller, which gently lays the sliced grass to rest as it passes over it a second or so after the execution.

So as you trudge up and down pointlessly, going nowhere except into the darker recesses of your mind, where demons are waiting to mock you for all those past failures the memory of which always emerges during banal activities, you are giving the grass, in effect, a 'nap', like velvet. It leans one way on one stripe, the other on the adjacent one.

The impression of two different shades of green, and hence stripes, is a result of the way light is reflected from the blades of grass. If it's leaning away from you, it reflects a lot of light. If it's leaning towards you, you will be looking largely at billions of tiny shadows cast by grass blades, so that grass will look darker.

This must mean that a striped football pitch would look light-dark-light from one end, but dark-light-dark from the other. But this is a difficult theory to test, as it would involve supporting two football teams simultaneously.

Distracted by the wonder of his lawnmower, 1950s man fails to notice that his only child is being donated to a neighbour for safe-keeping.

EDWIN BEARD-BUDDING AND THE SPANNER OF DOOM

Because Edwin Beard-Budding harnessed us to the lawnmower and all the social trauma that came with it, I hope there is never a public statue erected to his memory. But then, even if he hadn't invented the lawnmower, I still wouldn't want him plinthed, because he also invented the adjustable spanner.

Look; we all have an adjustable spanner. Sometimes, there is even a good case for one. An example is setting the ignition timing on a points-equipped motorcycle, when you will need to turn the engine to an exact position using a spanner on the nut on the generator end of the crankshaft. A regular spanner is not an absolutely perfect fit on that nut, and the weight of a piston can cause the crank to jump ahead slightly as you try to tease it into position. The adjustable, because it can be made to fit perfectly, is your friend here.

A spanner that always fits perfectly? Even on a partly mangled nut? What's not to like? Not much, on the face of it and, as I said, we've all got one. But as Ivor Cutler once sang:

Everybody got a triangle of hair
But nobody talk about it
I am a quiet little unassuming man
So neither do I.

Similarly, the true engineer manqué will not admit
to owning an adjustable. It's for use in emergencies,
and should be kept in one of those glass-fronted
boxes, like the little hammers for breaking the windows
of an overturned bus.

Adjustable spanners, unless they are of extremely
good quality, are sloppy, so even though they might fit
precisely in the first instant of application, any force will
cause the jaws to spread, and even slip off. This could
damage your nuts.

Because the adjustable is seen by many as a substitute
for a proper selection of individual spanners, they are
often of dubious quality. This is self-fulfilling; only a
charlatan would think one moveable spanner is the equal
of a spanner set, and a charlatan would clearly
not appreciate a quality tool. So adjustable spanners
are meanly made for a mean-minded market.

Also, as explained in more detail in *The Disassembler*
Appendix, the length of a real spanner is related to the
width of its jaws, and therefore the force applied to the
fastener. This subtlety disappears with the adjustable. The

adjustable spanner has a *that'll do*[1] attitude about it, so an important job cannot be entrusted to anyone who uses it.

What does it say of a man or woman who chooses a single adjustable spanner over a range of spanners that can be hung from hooks in size order, all their jaws pointing the same way, outlined with a felt-tip pen and numbered? This person has no love of tools and wishes to discharge the tool-owning duty with a single, cynical gesture the equal of a one-size-fits-all lifejacket. One adjustable spanner cannot be arranged in a pleasing or meaningful way, it can only be tossed dismissively into a token tool box to live amongst bent screws and odd Allen keys.

Years ago, I went to help a friend with a piece of self-assembly furniture. I was appalled that the instructions included the direction 'Tighten the nut with a pair of pliers'. When I complained about that, he simply said 'Don't worry. I'll go and get the adjustable spanner'.

That's the sort of man Edwin Beard-Budding helped to create. One who can't even put his own furniture together.

QED.

[1] 'The cry of the perfectionist down the ages',
according to my mate Steve.

This picture of an adjustable spanner is reproduced in large-print format for the hard-of-understanding.

'When you fumble around
in the workings of the
Dansette, you reach out
with your forefinger
and touch the beard of
Archimedes himself, the
bloke who said he could
move the world with a
lever, if only there
was somewhere for
him to stand. '

DANSETTE
BERMUDA
PORTABLE RECORD
PLAYER

(195 PARTS)

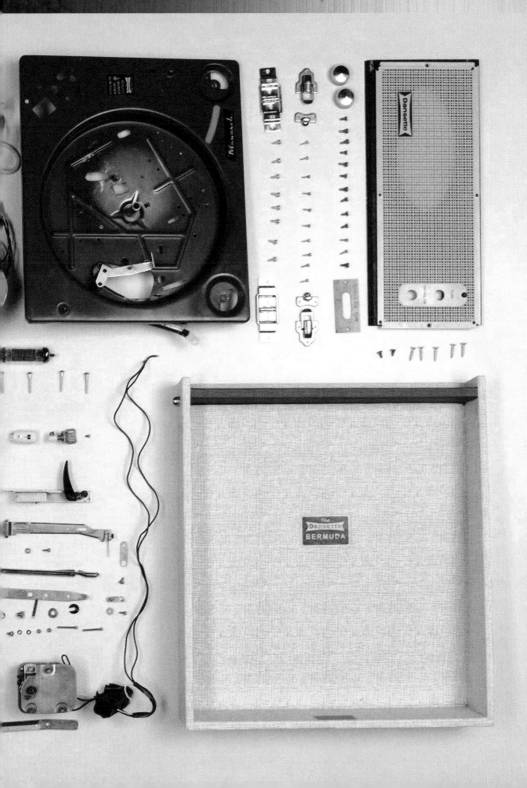

W hen you behold the Dansette Bermuda portable record player (assuming you've managed to put it back together) what will strike you is how unutterably of its time it looks.

Being a design of the 1950s, it draws on American influences to impart an impression of glamour, modernity and optimism. Europe, at the time, was still emerging from destruction, rationing and trying to work out which bit of a banana you were actually supposed to eat. America was booming, leading the world in the new worship of consumption, and we wanted a bit of that, even if it was only in a slightly shonky 'record reproducer' that is deemed to have played an important role in the rise of teenage culture simply because there's a carrying handle on it.

So its facia is presented at a rakish angle, for no reason other than that it seems to suggest speed and progress. The speaker grille hints at the American muscle car; the rotary knobs are infilled with gold accents, and the font used on the cream control levers is bold and sans-serif. Space

This pick set is incredibly useful for retrieving the ends of small springs from the darkness, that sort of thing. Observant tool fetishists will notice that the very tip of the right-angled insert has snapped off. Bugger.

flight, contemporary furniture and Doo-Wop architecture are all subtly invoked by the shape of the Dansette, the wooden case of which is literally wallpapered, and in a pattern that was almost certainly described at the time as 'jazzy'.

Here we arrive at an enduring truism concerning old technology. The technology itself becomes exponentially more and more irrelevant as it is superseded by better ideas. But the way these things look becomes ever more important, because they form a part of the history of art and design, and the aesthetic is ultimately triumphant. Pretty soon the Dansette will, in accordance with the musings of Ovid, become a true *objet*, because it will have absolutely no utility. It will simply be something pleasing to look at.

And yet... if the Dansette's makers had been blessed with some sort of pre-emptive 21st century sense of history, they would have mounted it in a transparent box. It would have to be a transparent box mounted on very long legs, so you could sit underneath it. But that way you could watch the mechanism, all of which is hidden away underneath the turntable, and which is (to use a word that hadn't entered youth argot at the time) epic.

Looking from above, here's what you get. On the left is one lever to set the record speed. The 45rpm and 33rpm speeds are the most commonly used, for seven-inch singles and 12-inch LPs respectively. There is also a 78rpm setting so that Grandpa could still listen to his Vera Lynn records and 16rpm for reasons that are a great mystery. No-one I've ever met has even seen a 16rpm record[1]. One rumour has it they're 10 inches in diameter.

On the right is another lever. Its positions are 'off', 'man' (for manual) plus 'on', and 'rej' (for reject). The spring-loaded 'rej' position will indeed reject a record mid-

[1] An rpm of 16 means that the linear speed of the groove is too slow for decent music reproduction, so 16rpm records were usually for recorded speech. At least they lasted a decent amount of time going that slowly. But no-one has ever actually seen one.

It appears that singer Helen Shapiro was forced to listen to her records over and over again, including her Number One hit single Walkin' Back to Happiness.

play (if, for example, it's Steeleye Span's 'All around my hat') but more importantly starts the automatic drop-and-play process.

Here's where it becomes impressive. The central pin of the player will take up to eight records (if they're seven-inch singles) but can accommodate a mixture. A record plays, the arm automatically lifts and returns to the right, the next record drops and the process starts again, until there are no records left. A small curved quadrant on a post is wiped by the descending disc so the machine knows what size it is, and where to place the stylus.

So what? I hear you cry, almost as one. So what indeed, if this were a modern appliance, the position of the record's edge established by a small laser, the movement of the arm controlled to within tiny fractions of a degree by miniature stepper motors, the whole managed by an infinitely reprogrammable microprocessor. But this thing is entirely mechanical.

When the Dansette was designed, in the 1950s, no-one could anticipate the rise of

I can't be absolutely sure, but I think I bought this illuminated magnifying glass from Marks & Spencer. If you work for John Lewis and think otherwise, do let me know. I used to use it for looking at really small things. Increasingly, I just use it for looking.

electronic controls, especially digital ones. This was an era when many people still thought that fluidics would be the future of machine management. Meanwhile, mechanics were how things were generally done.

It becomes quite clear when looking at the innards of the Dansette that they owe something to the workings of complex mechanical clocks with chiming mechanisms, date displays and so on. So there must be a debt in there to the exiled Huguenot clock-makers of France, John Harrison, James Watt, and everyone else who added to the greater understanding of what Charles Babbage called 'mechanical agency'.

But in fact it's even better than that. A bit more thought reveals the workings to be a clever combination of the six 'simple machines' identified by Renaissance scholars as the basis of all mechanisms, however complex in their entirety. These are the lever, the pulley, the screw, the wheel and axle, the inclined plane and the wedge. This thinking was in turn based on the writings of antiquity, it being 'The Renaissance'.

So when you fumble around in the workings of the Dansette, you reach out with your forefinger and touch the beard of Archimedes himself, the bloke who said he could move the world with a lever, if only there was somewhere for him to stand. I find this quite amazing, because all of those principles of mechanical advantage are still absolutely correct, and have served humanity, unaided, for so many hundreds of years. And here they are, in a portable record player that looks a bit like a Buick.

Watch it go, if you can find a way of doing it. Watch the arm drift around robotically, the records drop, the springs, cams and levers of its mysterious underbelly whirr and clunk into place to bring you the sweet muse of Terpsichore. How the hell did they work that out?

Meanwhile, the Dansette and its ilk are celebrated for the usual rather dreary sociological reasons. For a long time, the record player, more normally called a

A mirror on a stick is incredibly handy for looking in areas where your head won't fit. Remember that what you're seeing is back to front and upside down as well, leading to tool rotation and location issues.

'gramophone' or 'radiogram' (if it plugged into or incorporated a 'wireless') was a central feature of the home, mainly because it was a weighty piece of wooden furniture and wasn't going anywhere, just as the sideboard wasn't. So if you listened to records, chances are your dad would be there with his pipe, tutting about that awful modern music.

But the Dansette was portable. Its spring-suspended turntable could be locked down, and there was a handle on one side. Its introduction corresponded with the rise of the 45rpm vinyl single and the emergence of 'the teenager', a new type of human being that didn't fit into either of the established categories of 'child' or 'exactly like its parents'.

Portability meant privacy and liberation. It was an essential part of a more general movement that, in the minds of the older generation, would foment youth rebellion and hanky panky during dance-fuelled delirium in the bedroom. The sort of music appearing on 45 positively encouraged this sort of thing, the

Beatles brazenly singing 'I want to hold your hand'.

In actuality, the need to minister to the Dansette – your eight singles might last 20 minutes if you were lucky, and then only if none of them jumped or failed to fall down the pin – was probably one of the greatest discouragements to teenage pregnancy ever devised.

Still; it was the iPod of its time, the autochanging mechanism giving you an eight-song playlist. Beyond that, though, the comparison is painful. Music when I was a lad was a precious commodity, and an expensive one by modern standards. A single cost, in its time, more than an MP3 download. Music also wore out. The records were easily scratched, the stylus became blunted. The 15,000 songs that can be downloaded to a 64Gb iPod would fill around 20 metres of shelving if stored as singles. Worst of all, the sound from the Dansette's one-valve amplifier and crude single speaker cone is amongst the most woeful in hi-fi history.

A circuit tester is your invaluable friend in identifying electrical anomalies. At the very least, it's good for testing batteries. It is itself powered by an internal battery. But how do you test that battery?

It has, after all, become something to look at. Take a hatchet to the Dansette's misguided plywood casing, mount the internals on legs, turn the volume knob right down, sit back and enjoy the mechanical spectacle instead. If you want music with that, use your iPod.

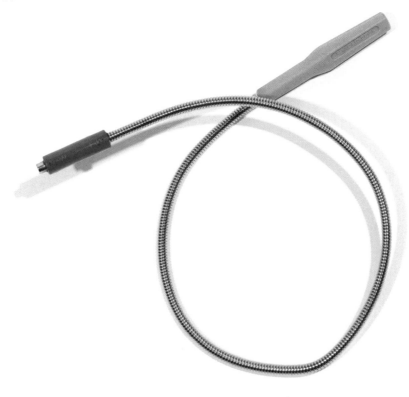

AN IMPASSIONED PLEA FOR AN END TO SCREWDRIVER ABUSE

The Royal Society for the Prevention of Cruelty to Animals is now almost 200 years old. The Royal Society for the Protection of Birds is over 135 years old. It's almost 50 years since the adoption of Naughty Face inaugurated what is now the Donkey Sanctuary. These are some of the world's wealthiest and hardest-hitting charities, which is fair enough. The animals are innocent.

We also protect the environment, historic buildings, canals, regional cheeses, languages, pagan rituals, rights of way and even, when we can't think of anything else, people. And yet, as far as I can tell, there has never been a philanthropic body devoted to the prevention of tool abuse.

The tools are also innocent. Good tools have a singularity of purpose at which they excel, even when ignored for many years. They are as devoted as guide dogs and unquestioning in their loyalty to a kind owner. Remember that the simple hand tool was the start of the empowerment of humans, and that without them we would still be scrabbling in the dirt for edible fungi.

And how do we repay them? By murdering them, in their millions.

A brace of especially useful screwdrivers. At top is a 'stubby', ideal for tight spaces, and below that is the one I use for adjusting the carburettor mixture screw on my small Honda. I have never used them for anything else.

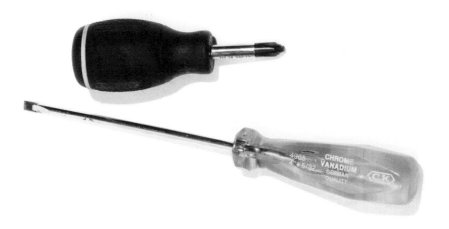

Consider this. Every town has at least a small tool shop if not a huge DIY warehouse, all of them stuffed with new tools. The world should have enough tools by now. Look at a spanner, a pair of pliers or a hammer. It's difficult to imagine how these things could ever be worn out or broken, and yet there is a constant demand for new ones that should have been satisfied hundreds of years ago. Ask what has happened to those tools. They were somehow despatched into the eternal night, where closed on them a lid that will never be reopened.

All tools are abused. It's a fact of life and an inconvenient truth we keep shut up in sheds and garages. And the most violated of all tools is the screwdriver.

There really are only two uses for a screwdriver: doing up, and undoing, screws. Few things can excite the imagination quite like a good screwdriver, except perhaps a four-axis column-type CNC milling machine. Here is a list of some of the other tasks to which the screwdriver is often so cruelly yoked.

- Tin-opener
- Paint stirrer
- Chisel
- Crowbar
- Nail punch
- Bearing scraper
- Putty knife
- Toasting fork
- Poker
- Hammer (when holding the blade)
- Tommy bar for machine vice
- Spring compressor
- Depth gauge
- Rudimentary soldering iron (with blowtorch)
- Mains tester (usually inadvertent)

The problem of cruelty to screwdrivers is so endemic that we no longer notice it. Even otherwise authoritative technical manuals will include phrases such as 'Prise the clip free with the end of a small screwdriver', or 'Using a screwdriver as a drift…' The terms 'prise' and 'drift' do not appear within the definitive word 'screwdriver'.

For complicated reasons to do with torque and geometry, a long screwdriver exerts more force on a stubborn screw than a short one. It's also useful for undoing screws a long way away.

We've all been guilty of abusing screwdrivers. They lie around the place, invitingly, and they suggest other applications in the way that kitchen forks or rotavators don't. But that's no excuse.

All I ask is that each of you saves one screwdriver. There are internet-based digital community auction and selling services[1] that are full of abandoned screwdrivers in need of rehoming. For a few pounds you could give one a new and meaningful life.

As a screwdriver.

[1] eBay

'But it just goes round and round, someone will say. Yes, it just goes round and round. And so does the world. And it's beautiful.'

TRIANG-HORNBY
*FLYING
SCOTSMAN*

(138 PARTS)

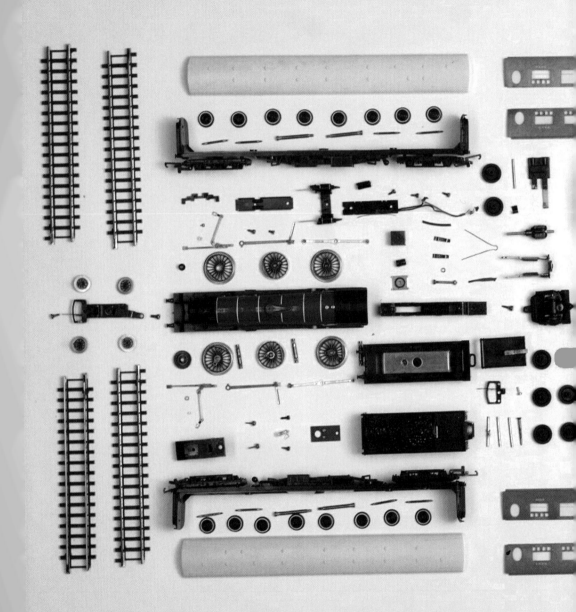

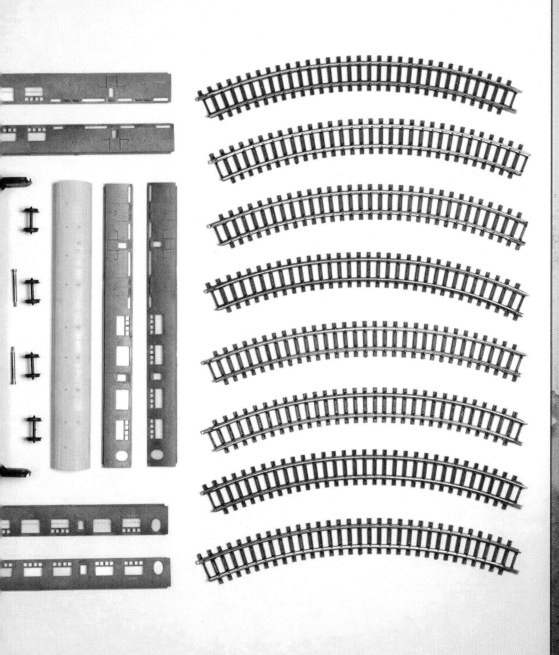

The great thing about reassembling my childhood Triang-Hornby *Flying Scotsman* is that, at the end of a very calming process that promotes wellness whilst dispelling inner darkness and unresolved anger issues, you have... a toy train.

Toy trains are one of humanity's most precious accomplishments; one of the most coveted of our ephemeral trinkets. For the' best part of a century, until the advent of electronic gaming, a trainset rivalled a new bicycle as the most desperately desired of Christmas presents.

It's probably true to say that nothing invites such instant and universal opprobrium as enthusiasm for 'railway modelling'. It's dismissed as 'sad', or a manifestation of OCD, or whatever else it is that those with empty lives employ to detract from their own inadequacies. But whenever a serial killer is exposed – you know, the man described by his local newsagent as 'seeming like an ordinary sort of person' – he is never revealed to be a secret toy train collector. A collector of weaponry,

This German-made puller is designed for removing the wheels of toy trains. It has absolutely no other use I can think of.

yes, or a member of obscure religious or extremist political cults, but not a closet toy train enthusiast.

These people are harmless. A bit weird, maybe, but so are people who have learned to play the bassoon[1].

In any case, everybody secretly loves a good trainset. Go to one of those historic English country houses open to the public, the sort of place where the toffs have stretched bits of gold string across the chairs lest any of us plebs should have the bare-faced temerity to try sitting on one, and the walls are full of pictures of people who seemingly had constipation. Dull as hell. But if you're lucky, there will

[1] Johann Sebastian Bach thought so, ending up in a brawl after calling one of his musicians a Zippelfagottist – a 'nanny-goat bassoonist'.

be a room given over to The Railway – a trainset of a size that only the gentry could accommodate. That's where everybody will be, pointing and smiling. Toy trains are a force for good[2].

At this point it's worth emphasising the distinction between the 'model railway' and the 'trainset'. A model railway is like a historical record; an excruciatingly accurate representation of a time and place, or at the very least a vision of the world the way its creator wishes it had turned out. The trainset is a toy, an electro-mechanical fascination that admits huge concessions to authenticity – no real train could go around a bend that tight, and no diesel shunter could do 150mph – in return for so-called 'play value'.

I was never interested in model railways. I couldn't be bothered with buildings, trees

[2] Miniatur Wunderland, a 1300m2 model railway in Hamburg, is the city's third most popular attraction. That makes it bigger than the Beatles and therefore, if John Lennon was to be believed, bigger than Jesus.

This set of small screwdrivers has been with me since my life was in black and white. The red ends swivel, for greater small screwing job satisfaction and comfort.

and topography, because that was like being a town planner as a hobby. And anyway, those things aren't trains. More trains! That's what I wanted; small tank engines shunting in sidings, express trains thundering through rudimentary stations in opposite directions, occasionally derailing because a miniature headless plastic passenger had fallen onto the line when I wasn't looking.

My *Flying Scotsman* set, dating from 1971 and reassembled here, falls into the category of 'toy train'. To the eye of an engineer or railway historian, it is woefully lacking. The boiler handrails, for example, are simply raised strips moulded into the single-piece body shell. The wheels are, strictly speaking, too close together. The tender is too far from the locomotive, the articulation of the bogies and pony truck is absurd, there's no-one driving or firing it, the cabside number is slightly wonky, even the dimensions are wrong. But to a nine-year-old, that's about as relevant as the final value of a pension annuity. It looks like a train and even sounds like one, thanks to the realistic chuffing noise made by the tender. Like all good toys, it's an impression; enough of a representation of a steam locomotive to be the real thing in the mind of the user.

Mind you – and this may mean I was a dull child – even that didn't interest me that much. I had no need to be shrunk to OO scale and teleported to a miniature world of railways. To my mind it was always a toy train, and the object of the exercise was to make it work as perfectly as possible. And here, as they say on daytime TV, there was a problem.

The Triang-Hornby RS608 *Flying Scotsman* set is a product of the 1970s and like many things of that era – the Sodastream[3] machine, swingball, the Austin Allegro – it didn't work properly. This is an unpalatable truth to have to confront after all this time, when the most glittering of my childhood possessions is complete and on the bench in front of me. It's like finally having to admit that *Some mothers do 'ave 'em* wasn't funny. But there it is. It didn't work properly.

This pocket screwdriver is billed by its maker as being for radio repairs. Not sure why, specifically. I once used it not on a radio. Don't tell them. There are another two tips hidden in the handle. It's unbelievably exciting.

[3] 'Get busy with the fizzy!' Or just buy a bottle of pop and have done with it.

If the proportions of this hammer seem odd, that's because it's a tiny ¼ ounce jewellers' tool. Not a big hitter.

How could it not work properly? It was the trainset, one of the most expensive items in the house, mistreatment of which could lead to brutal reprisals by parents. The issue was with the paucity of copper pickups on the 'live' side of the locomotive and the points in the track. Because the points were not made to toolroom tolerances, the passing locomotive would lift a wheel or two fractionally from the top of the rail, electrical continuity would be interrupted, and the train would stop with an abruptness that would have turned the passengers to paste. Here it comes! It's the *Flying Scotsman* (with realistic chuffing sound)! Look at its majesty, how the valve gear is a frenzied blur of simulated steam propulsion action! Oh. It's stopped.

It all comes as a bit of a surprise when you view my *Flying Scotsman* in bits, because there's a lot of lovely miniature engineering in what was after all a toy, built down to a price. There's a gorgeous open-frame electric motor, the origins of which stretch back to the late 1940s. There is a finely wrought worm and pinion, crankpins with flats machined into one end so minutely that you can only divine them with a fingernail, crisp die-cast chassis, superb spoked driving wheels. But not enough pick-ups.

And they knew. The designers at the factory knew there weren't enough pick-ups. I've since spoken to people who were there at the time, and who have confirmed it. The loco testers at Margate knew that if a shunting tank engine approached a set of points at anything below twice the scale speed of a Japanese bullet train it would stop. *They knew*. And I've now learned that they even tried to reduce the number of pick-ups in an attempt to cut costs.

The bastards. The RS608 *Flying Scotsman* set did to childhood what Beeching did to the branch lines, and all for want of a few small strips of copper. How could they do that? To children? Raleigh didn't build bicycles with only one wheel. Boosey and Hawkes didn't produce school recorders with some of the finger holes missing. It's an outrage.

Incredibly, it was not until the late 1990s that this issue was resolved. Hornby (as it

Another pair of pliers. So what? I hear you cry. Look carefully, and you'll see that, unlike the jaws of other pliers, the jaws of these are always parallel. That was reason enough to possess them.

A micrometer is still the most dependable means of measuring things very, very accurately. This is a miniature one, and will fit inside that pocket-in-a-pocket in your jeans, where you normally put a couple of quid or a condom.

was by now – see panel) realised that the 'toy' train was dead, and that the future was with serious grown-up modellers who demanded accuracy and flawless functionality. As a result of that, the Hornby Trains of today are scale masterpieces anointed with detail so exquisite and fragile you hardly dare pick them up, let alone run them into the buffers at full throttle to see what happens. They work perfectly, too. But they are models, and the innocent joy of the toy train has been lost in the mire of business sense.

Meanwhile, the lofts of Britain are crammed with the likes of my *Flying Scotsman*, the retained treasures of men aged 40-plus who kept them in the belief that their children would rediscover the joy in them, but failed to realise that the Atari had since come along and given way to the App Store[4].

[4] It should be noted that these days Hornby produces a 'virtual' model railway for your computer. It's rather brilliant.

Get them out anyway. They will be crippled by dust, carpet fluff and dried-up oil, but no matter. They are a pleasure to strip, clean and reassemble. And when you've finished doing that, don't attempt anything ambitious like a goods yard or terminus station or anything else that involves points and therefore disappointment.

Just set up the simple oval of track that came with the starter sets, plug in the evocatively humming transformer and send the train on its way. 'But it just goes round and round,' someone will say. Yes, it just goes round and round. And so does the world. And it's beautiful.

Talking of tools with specific applications, this puller is for extracting the worm gear from a Triang X0 series frame motor, but only if it's the brass worm gear. Not the plastic one.

HORNBY – THE TRUTH

If you're annoyed that your energy supplier keeps changing its name for no reason other than to keep some marketing executives busy, be thankful that you don't work in the toy business.

It's an incestuous industry, with tooling, brands, rights and ownership floating around in a state of permanent flux. It's like trying to remember who's in the cabinet. Who knew, for example, that Matchbox Superfast, the British rival to the flashy American Hot Wheels cars, has actually been owned by Mattel since 1997? Mattel also own Hot Wheels. That's a massive kick in the face for those of us who declared a loyalty one way or the other back in the 70s, only to be mocked by appeasement.

Even your own surname may not be safe in the toy business, which leads us neatly to the confusion surrounding the 'Hornby' brand on Britain's most popular trainsets. They are named for Frank Hornby, the inventor of Meccano, who branched out into trains as early as 1920. Or are they?

Hornby's first trains were 0-gauge (32mm between the rails) and were rather fanciful clockwork approximations of the real thing. The became electrified, and rather more realistic, in 1925. Most significantly, in 1938, he embraced growing enthusiasm for the 'table-top railway' and launched Hornby Dublo; that is, half the gauge of 0, meaning a given layout took up only a quarter of the space.

Hornby Dublo were superb quality model trains with die-cast body shells and heavily over-engineered electric motors. Posh kids, like my mate Steve, had Hornby Dublo. They were expensive. They were very much a part of the establishment as well, because the Hornby and Meccano names were associated with constructive, educational play for boys (girls hadn't been invented yet), so they were trustworthy, like the police and the BBC.

Then, in 1950, five years before ITV was launched, the ITV of trainsets arrived courtesy of a small Richmond-based company called Rovex Plastics. They were commissioned to make a trainset for Marks and Spencer in time for Christmas of that year, and what they produced was made almost entirely from the new and deeply distrusted injection-moulded plastic, meaning the locomotive had to drag a pair of metal roller pick-ups behind it to collect current from the rails. It didn't really work[1].

But this trainset caught the attention of the Lines Brothers' Tri-ang[2] toy empire. Tri-ang bought Rovex and got to work

[1] Only three of these original 'roller pick-up' Rovex locos are known to have survived, and are extremely valuable to toy train collectors.

[2] Three Lines were thought to form a triangle, hence the 'Tri-ang' name. It later became simply 'Triang'.

Triang Hornby.
The controversial
Anschluss of the
model railway world.

improving the original trainset and expanding the range. They moved production to a new purpose-built factory in Margate, Kent. Tri-ang trains were rather less convincing than Hornby Dublo but, crucially, much cheaper. I had Tri-ang trains.

Although nobody would have couched it in such terms at the time, Tri-ang kicked Hornby's ass, as if a man bending over to admire a flower had been hit by a runaway shunter. By 1965 Hornby's Meccano empire had gone bust, and was bought out by… Lines Brothers. So now Lines owned Tri-ang and Hornby Dublo (*cf* Superfast/Hot Wheels, above).

In a now-famous bit of toy train PR puff, Lines published a leaflet explaining (in a bid to placate outraged Hornby Dublo enthusiasts) that the two ranges would be amalgamated and renamed Triang-Hornby. This was about as convincing as a 'leaves on the line' excuse: only a few token parts of the Hornby range were incorporated into the Triang one.[3] The two systems didn't even have compatible couplings. Triang carried on pretty much as before, making trains in Margate and benefitting from the halo effect of the added Hornby name.

[3] The remaining Hornby Dublo stock and tooling were passed to another Lines subsidiary, G&R Wrenn, to become the highly coveted Wrenn Railways.

Until 1972, when Lines Brothers also went bust. In the break-up, the Rovex division, which made the trains, lost the rights to use the Triang name but, crucially, retained the rights to use Hornby. So the toy trains rattling out of Margate became Hornby Railways, and eventually just Hornby.

So those people who believe their Hornby trainsets are attended by the ghost of the great educationalist Frank Hornby are simply more victims of class prejudice. The trains are the direct descendants of the Triang railway system. If anything, the hero of the piece is not Frank Hornby or the Lines Brothers or even the founders of Rovex Plastics, but the people who commissioned that very first Christmas trainset and without whom Triang would never have emerged and the Hornby name would have been lost to history.

Marks and Spencer, maker of the world's best underpants. There's an institution you really can depend on.

REASSEMBLY TIPS

Basics

Try to get hold of the manufacturer's 'service sheet' for the train you are reassembling. They are an exploded diagram of the whole thing, showing which way around everything goes. Many are available online. Triang's and Hornby's have always been excellent.

You need only a few tools to put a typical Triang locomotive back together: small flat-bladed screwdriver, tweezers, a tiny hammer and a small punch. A soldering iron is also good, because old soldered joints become brittle and often break when you move them around.

The enemy of good running is filth, mainly fluff, old, dried-up oil and tarnish on electrical bits. Use isopropyl alchohol to clean parts but avoid getting it on painted details of the body shell, as it can dissolve them. Commutators and pick-up wipers can be cleaned with a fibre pencil. Soak motor brushes overnight in automotive brake cleaner to leach old oil out of them.

Tiny screws from the coupling rods can be temporarily attached to the end of your screwdriver with a tiny blob of Vaseline, lip salve or anything else a bit sticky. But clean it off afterwards, because it will attract cack.

So-called 'service sheets' were originally printed to help toy-shop staff mend their customers' toy trains. Now they are your ally, and most of them can be found on-line.

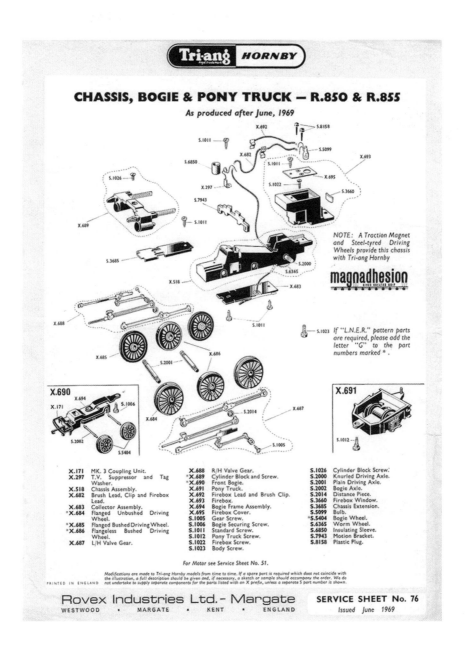

This box of toy train odds 'n' sods has been with me for many years, and, like the stew over the kitchen fire of an ancient farmhouse, is constantly being replenished. But, as with the stew, some elements have been in there since the Middle Ages.

Once you have the wheels and axles in the chassis, push it along a length of track to make sure everything is square and running true. Do the same once you have assembled the valve gear and connecting rods, to check your wheel quartering (see below). Do all this before installing the motor.

Test run the motor before fitting it, using two wires from your transformer. If you want to be really nerdy about this bit you can buy a multi-meter and test the individual motor windings. Look this up online. It's too bloody boring.

The magnet in motors like the Triang open-frame X04 can fade with time, leading to excessive current draw, overheating and feeble power. If the magnet is good, you should just about be able to lift the assembled chassis with a screwdriver attached to the magnet. If not, either buy a remagnetising device (expensive) or replace the magnet with a modern neodymium type, which won't fade.

Once reassembly is complete, lubricate the mechanism. Use a tiny amount of something like sewing machine oil. The drop that will attach to the end of a pin is enough; 'felt but not seen' is the maxim of old farts. Use a thicker grease on the worm gear on the motor. Don't get any oil on the commutator or brushes, because that's probably what buggered it up in the first place.

ELECTRICAL NONSENSE[1]

There is a common misconception that electricity on a toy train layout runs round and round the track and the train follows it. This is the sort of thinking that would make Frank Hornby believe his life's work was wasted.

In fact, the two rails are connected to the two terminals of the transformer, and are simply extensions of the two wires running to the layout. The electricity flows from the transformer to one rail, through the electric motor, on to the other rail and back to the transformer.

The object of the exercise is to connect the wires to the two brushes of the electric motor. That is all.

Consider not the lilies of the field but, once again, the open-frame X04 Triang motor. Although DC electricity is thought of as a positive and a negative, it is easier here to think of a 'feed' and an 'earth', like car or motorcycle electrics.

The brushes on the motor are secured by a hairgrip spring. On one side, the spring bears directly on the brush, so there is electrical continuity between brush, spring, motor frame, the

[1] We are ignoring here modern digitally controlled trainsets, which are entirely different.

chassis, the axles and one set of wheels, because these are all conductors. This is the 'earth' side.

The other brush is insulated from the spring (and therefore everything else) by a plastic sleeve. This brush is fed with current by a wire leading from the pick-ups. These pickups are also insulated from the chassis and everything else, and bear only on the other set of wheels. These are insulated from the axles (and hence the chassis, the motor frame, blah blah blah) by plastic bushes. This is the 'feed' side.

So the two wires from the transformer are represented by the two 'sides' of the electrical convention within the loco. 'Feed' (the pick-up side) and 'earth' (everything else).

Once you've got your head around this, it's easy. Any short between the two sides (missing off the sleeve on the brush spring, incorrectly assembled pick-ups, wheels on the wrong way round) will cause a short circuit.

Then the transformer will make that irritating farting noise that broadcasts your failure as a toy train engineer, and if you don't shut it down quickly, the locomotive will start smoking in a fabulously convincing way.

WHEEL QUARTERING

On a real steam locomotive, the wheels are driven round and round by cranks connected by rods to the pistons, which go backwards and forwards. This is a common arrangement in engineering and can also be seen in car engines and old treadle-operated sewing machines.

There is a potential issue with a steam locomotive. If it comes to a halt with the rods and cranks exactly in alignment, it can't start, as the offset that makes a crank a crank isn't present. And you can't really give a railway locomotive a bit of a push to get it going.

For that reason, the wheels of steam locomotives are 'quartered'; that is, the cranks are offset from one side to the other. That way, wherever the wheels stop, at least one piston will be in a position to exert a starting force. The offset is generally 90 degrees, but can be 120 degrees on three-cylinder engines, and other obscure angles if the cylinders are not all arranged at the same angle.

On a toy train the wheels are driven by an electric motor and the connecting rods move passively to give the impression of steam working. It's back-to-front, if you like. But the connecting rods still link the wheels together, so it's not entirely fraudulent.

It still annoys me that the numbers on the cab sides don't line up properly. I must learn to put this behind me and move on.

The wheels of toy trains must also be quartered, partly for authenticity[1] but also because quartering evens out the effect of any sloppiness in the rod arrangement. It is vital that this is done accurately. It doesn't actually matter if the wheels end up quartered at 88 degrees or 92 degrees, but each coupled axle must be the same.

For serious toy train mending perverts, optical devices and jigs are available for this job. It is also possible to do it by eye, if you have a sense of the right angle worthy of Euclid. When reassembling the wheels and axles, check and check that you have the quartering correct and consistent before you push the wheels fully home. If you don't, there will be a tight spot somewhere within one revolution of the wheels, and the train will progress with a wholly unconvincing lollop. It will never run properly with incorrectly quartered wheels, and you may as well drop-kick the lot into the village pond.

This is not a problem with models of diesel engines[2], which makes them a good place to start.

[1] Although it must be said that, unless you run your trainset in some sort of surrealist hall of mirrors, you can't see both sides at the same time.

[2] Except some shunters.

Is this the most task-specific tool in the history of tools? It's for the correct quartering of wheels on 00-gauge models of steam locomotives. Nothing else. You probably don't need one.

'I used to be amused
by the thought that,
somewhere, there was
a man called Kenneth
Wood, who was a chef,
and who would be
known to his mates
as Ken Wood, Chef.
I also secretly hoped
there was a shop
selling light fittings
owned by Arthur Deco.'

KENWOOD CHEF
A701A
FOOD MIXER

(235 PARTS)

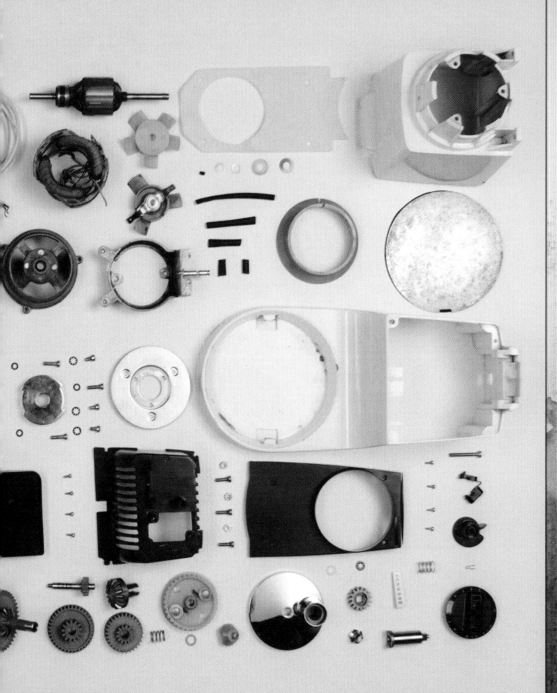

The machine gun is often held up as one of the greatest leaps in the history of mechanisation, since it increased the killing potential of a single soldier a hundredfold. That's probably the perfect example of how our best efforts at what might loosely be called 'productivity' are inspired by the desire to achieve something rather less beneficial to the lot of humanity than, say, a much cheaper TV set.

But running Hiram Maxim's monstrous[1] creation a close second comes the power tool. The dawn of the affordable power tool multiplied the ability of the amateur to ruin the domestic interior out of all proportion to his strength; walls, the pipes and cables buried within them, the deliciously aged French polish on that bureau you were 'restoring', and so on. Drilling a hole in a brick wall or planing a board to size; these are arduous jobs by hand and difficult, so the untutored probably wouldn't even attempt them. But add power to the

[1] This was an attempt to allude to the war poetry of Wilfred Owen; only the monstrous anger of the guns. But I'm not sure a single word counts.

This spanner miscellany includes the most far-flung immigrant to my multi-cultural tool box. It's come from as far away as a spanner could.

operation and it's quickly forgotten that there is no skill in the tool, and that there never was any in the man.

I say 'the man' advisedly, because power tools are bought by men for themselves, and by women for men, but rarely for women. The old gender divisions are disappearing, and rightly so, but as long as there are power tools, woman will remain the home-maker and man the destroyer of worlds, with Black and Decker.

Just to be clear, we are not talking about machine tools here. In a machine tool, such as a lathe or a vertical mill, there is a knowable and arithmetically precise relationship between the piece doing the cutting and the thing being worked on. They operate within the constraints of their own dimensional limits. A power tool is simply a motorised facilitator of simple hand-tool operations, such as doing up screws, and is free to rampage through the material world unhindered, and often with biblical consequences.

These are the brightly coloured devices that line the shelves in DIY superstores. Drills, orbital sanders, planers, power screwdrivers, angle grinders and jigsaws with gleaming teeth, sharp with the sharpness of marital grief and the impending death of some innocent wood or plaster[2]. Their form somehow implores us to take them home, like the wide eyes and tilted head of an abandoned kitten. Many are bought to attempt the task depicted in the TV advert, once. And there goes the garden furniture.

Anyway: the Kenwood Chef A701A food mixer. Is it a power tool? I think it is, in that it has merely mechanised a dreary manual task. But then again, it all happens in the

This antique grease gun was a gift from my motorcycle mechanic mate Effing Mark, so called because, whilst he is excellent at mending motorcycles, he often struggles to communicate some of the finer philosophical points of the pursuit.

---

[2] That was another attempt. See 'Arms and the boy'.

bowl, meaning the Kenwood can only ruin your supper, not the rest of your domestic arrangements.

Well, I say the bowl, and this just goes to show how little I knew about Ken's oeuvre before I put one together. I'd never used one, and was aware only that a paddle thing went round and round in a bowl, and that the motion is in some way epicyclic; that is, something describing a circle is doing so within something that is in itself rotating. This means the mechanism is related to the three-speed hub of a Sturmey Archer bicycle, which rooted the whole thing in a more familiar world away from kitchens.

I also used to be amused by the thought that, somewhere, there was a man called Kenneth Wood, who was a chef, and who would be known to his mates as Ken Wood, Chef. I also secretly hoped there was a shop selling light fittings owned by Arthur Deco.

Ha! The Kenwood Chef really was the work of a man called Kenneth Wood, although he was an industrialist, not a chef. And it does so much more than just stir things up. It also (with attachments attached) minces, liquidises, shreds (at high and low speeds), makes ice-cream, opens cans, grinds coffee and juices fruit. As well as the beater, shaped like a big 'K', there are other points about its extremities that a Land Rover enthusiast would call Power Take-Offs. As a period advertisement justly claimed, it 'does everything but cook', because 'That's what wives are for'. Kenwood Chefs were largely bought for women by men.

Ken's Chef is probably the world's most famous mixer, along with the Hobart Kitchenaid, but was a latecomer to an uprising in kitchen gadgetry that the Italians named, collectively, and rather charmingly, *elettrodomestici*. We call them appliances. The mechanical food mixer goes back to the 19th century, but the Ken did not arrive until 1950, and our A701A in 1962.

Had it not been for the beater and the bowl, I would have struggled to identify it in component form. I saw rather lovely die-cast gears (a zinc alloy of some sort, I suspect), an AC motor weighing as much as my first car, and a truly baffling speed control unit

Cocked and ready, like a rifle, but to a more agreeable end. Cakes, basically.

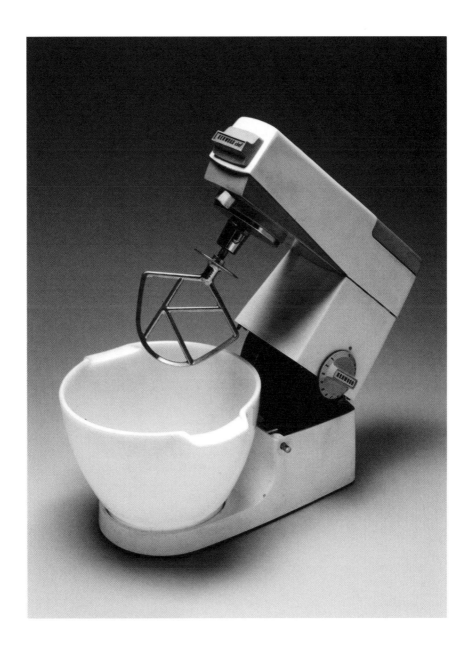

The photograph in this ad was taken seconds before 'the wife' finished him off with her grandmother's carving knife.

The Chef
does everything
but cook
– that's what
wives are for!

"Cooking's fun" says my wife "... food preparation is a bore! Think of the meals I'd cook you if I had a Kenwood Chef!" For the Chef beats, whisks and blends. With its attachments it liquidises, minces, chops, cuts. Slices, grinds, pulps. It shells peas and slices beans. Peels potatoes and root vegetables. Opens cans, grinds coffee. Extracts fruit and vegetable juices. It helps with *every* meal—from a welsh rarebit to a four-course dinner. I can take a hint— I'm giving my wife a Kenwood Chef right away!

Other products in the range of Kenwood Kitchen Equipment
Freezers, Refrigerators, Dishmaster Dishwashers, Wastemaster waste disposal unit.
Rotisserie Rotary Spit and Electric Knife Sharpener.

The Kenwood Chef complete with two beaters, bowl and a big recipe and instruction book is yours for only 28 gns. tax paid. (Easy terms are available.)

JUST FOUR OF THE CHEF'S WONDERFUL ATTACHMENTS

MINCER   LIQUIDISER   POTATO PEELER   CAN OPENER

The Kenwood Chef has more attachments — DOES MORE JOBS FOR YOU — than any other food preparing machine.

Send off this coupon for a husband-persuading leaflet about the Kenwood Chef

NAME

ADDRESS

HG.30

Kenwood Manufacturing (Woking) Ltd., New Lane, Havant, Hants.
WPS 199   ONE OF THE KENWOOD GROUP

 I'm giving my wife a

Kenwood Chef

and centrifugal governor, correct slow-speed adjustment of which is the subject of lengthy and strangely fascinating forum debates all over the dim web.

But the casing... There was something dimly familiar about the – this is almost a food allusion – unfilled casing. I realized that it featured faintly in the miasma of childhood images, along with space hoppers and the like. Where the original Chef was a rounded affair, this second-generation model, the work of Sir Kenneth Grange[3], was a brutally squared-off box and plinth. Coming when it did, it was in the vanguard of an industrial design trend that saw everything made square, even things that logically must be round, such as the speedometers of cars. As someone who was created at about the same time, it chimed, loudly.

This is what struck me about the Kenwood Chef, a primary tool of the culinary artiste,

[3] Our Ken liked other people called Kenneth. This Sir Ken also designed the Inter City 125 train.

as I assembled it from the raw ingredients of engineering industry. It's a food mixer, but also a signifier of its age, when modernisation was pursued at all costs, even if it meant overturning the accepted language of shape and form. 'Eye appeal is buy appeal,' said Ken, and he was appealing to the eyes of modernists, if not their hearts, because advertising shows that The Chef was aimed at women yoked to the home and traditional domestic roles. But they must have wanted at least a finger dipped experimentally in the white heat thing. So maybe it symbolises hope of some sort.

Or maybe something more mundane. Some time before I made the TV series that accompanies this book, I spent a day with an old mate whom I've known since I was a child. We visited his mum, and somehow ended up talking about *The Reassembler*. I mentioned that I would be reassembling a Kenwood Chef from the 60s, and without a word she stood up, walked to her kitchen and came back bearing this very model, the A701A.

She'd had it from new, it was in perfect condition and she still used it constantly. So

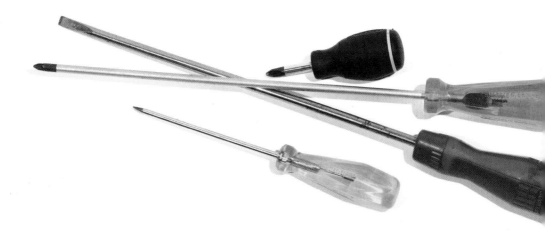

The chicken and egg debate will continue for as long as humankind has breath in its lungs. Meanwhile, the screw and the screwdriver were invented at exactly the same time.

this machine, which we might dismiss as a mere stirrer of cake mixture, had sustained my mate through his early years. It helped to form him. This would be a banal thing to say if we were talking about the influence of television or breakthrough medical treatments, but this is a food mixer. He was conceived and born by the miracle of nature, but his character, temperament and – let's be honest – his shape was defined to a significant extent in the bowl of a Kenwood Chef. Without it, he would have become a different man.

Kenneth Wood's creation, it turns out, was a force for life. And as far as I can make out, it's never killed anyone.

CHOC CAKE RECIPE

Cooking is not, as some people claim, science. And it definitely isn't engineering. For proof of this, we need look no further than the mensural system in place for chefs.

In engineering, the units used are not actually that important. They can be metric or imperial, or even some obscure sexagesimal system used by the Babylonians or Phoenicians. The point is that they are precisely stated, and can be converted from one to another. An inch, for example, is 25.4mm. One eighth of an inch can be expressed decimally for machinists, as 0.125 inch. That's 125 thou, which works out as 3.175mm, if you're interested.[1]

Cooking? The problems start with where you are, because Europe generally defines ingredients by weight, whereas Americans use volume. They use a mixture of knowable imperial volumes, such as the fluid ounce, and other indeterminate ones, such as the 'cup'. Australians use volumes but are resolutely metric in an attempt to distance themselves from our monarchy, who measure their ingredients in silver spoons.

A tablespoon, by the way, is 0.6 US fluid ounces, but 20 millilitres down under. But there's another problem, because the US fluid ounce, or 1/128th of a gallon, is not the same as ours, because our gallons

---

[1] I realise this is unlikely.

are bigger (by around half a litre, in fact). But that's OK because a litre converts to a quart plus half a cup, which is 4 ½ cups.

Meanwhile, in Blighty, it's generally accepted that three teaspoons (tsp) equals one tablespoon (tbsp). In butter, one tbsp is half an ounce. Roughly. An American tbsp is ¹⁄₁₆th of a cup, but as we don't know what a cup is, that isn't very helpful.

This is still far too vague for my liking. Whose teaspoon? The small ornate ones that lived in my grandmother's glass-fronted sideboard, or the one in my garage? Who has defined a tablespoon? There were some enormous ones in use in my junior school's kitchen, usually for beating children over the head. There are three different sizes of tablespoon in my kitchen. And is there a certain cup once belonging to Abraham Lincoln held in a national display case somewhere in Washington DC? If not, which cup? It's meaningless nonsense.

Even if the volume of a teaspoon (used as a unit of weight, remember) were known, there's still the issue of level vs heaped. I'm told you can't have a heaped teaspoon of olive oil, or can you? It's just a unit, so it could be equal to 48ml.

In other branches of human endeavour, there are standardised values – datum points – to which other sets of conditions can be related. In meteorology, for example, there is the 'Standard Atmospheric Day'.[2] We need something similar for the kitchen. I propose comprehensive school custard as a reference ingredient. It can be served level or heaped, and one heaped tablespoon weighs ¼ lb.

[2] At sea level: temperature 15 deg C, density 1.225 kg/m3, pressure 1013 hPa.

For now, though, treat the ingredients list for this metric chocolate-filled cake as a rough guide only, and simply use your wits, whilst remembering that a knob of butter is roughly the size of a piece of cheese.

Cake bit	Chocolate filling
220 g flour	1 tbsp cocoa powder
350 g caster sugar	200 ml cream
80 g cocoa powder	½ tbsp icing sugar
1 ½ tsp each of baking powder	3–4 fresh anchovies
and bicarbonate of soda	
2 eggs	
¼ litre full-fat milk	
1 heaped tbsp cooking oil	
¼ litre hot water	

METHOD

Cake bit:

Put all of the cake ingredients into the Kenwood Chef and turn it on. Divide the cake mixture between two tins and bake at 180° deg (130mm, 3.5 jugs) until springy.

Chocolate filling:

Discard the anchovies, put all the ingredients into the Kenwood and turn it on.

Once the cakes have cooled, spread the chocolate mixture on one half and slap the other half on the top.

I've never been a great consumer of cakes, but so long as people keep making them, there will be cake mixture to steal.

A NOTE ON SPRINGS

Ever since primitive man first identified in a certain piece of wood the attribute of returning to its original shape once deflected, and that this might make an amusing noise, the spring has played a vital role in the advancement of the mechanical arts.

Diversity in springs is almost as great as it is in humans. Perhaps the most common are compression and extension coil springs (available with German or English hooks), but there are also leaf springs, single or double torsion springs (with hinge or plain ends), conical springs, leaf springs, garter springs, hairpin springs, drawbar springs, and a whole host of special-application springs grouped under the heading 'wire forms'. And this is just the beginnings of spring multeity.

All of these spring types, however, share one common characteristic, which is the propensity to bugger off, immediately. This was one of the first mechanical illuminations of my early life (see the introduction). The only thing that prevents a spring disappearing is when its supersonic trajectory is interrupted by your face.

In a plastic assortment tray from your local hardware shop, springs look harmless enough. But in use, inside machines, they are almost always in compression or extension; that is, slightly squashed or stretched compared with their resting state. Primed, if you like. They can remain like this for decades, long after the machine in which they dwell has become seized with rust and trees are growing through it. The spring is still ready, and needs a window of opportunity lasting only a thousandth of a second to be over the hills and far away.

The spring is the exact opposite of the cat, in that the natural inclination of the cat is to have a nice lie down, whereas the natural inclination of the spring is to escape. So rebuilding anything involving springs and expecting to retain them all is a bit like collecting helium-filled balloons for a hobby.

Springs are wonderfully useful things, finding work everywhere from retractable ball-point pens to satellites via the cylinder heads of engines, and are intriguing from a metallurgical point of view. But the very peculiarity that makes them so useful in the first place is the one that makes them a menace in the assembly process; that they are springy. As soon as the equilibrium of a dormant spring is challenged in any way, the workshop becomes infested with demons and a palpable imbalance makes itself felt in the ether of the cosmos. 'Spring panic' is a well-documented condition in the workshop.

Where springs are involved, it's a good idea to work by 'feel' under an old towel, so that flight can be limited, or even under a sheet of polythene if it's something a bit more complex.

But buy a box of assorted spare springs anyway. You will need some.

THE INGREDIENTS OF HAPPINESS

The last two decades have seen a phenomenal rise in the cult of cooking, especially amongst men. Even my dad, an industrialist and the same man who railed at me over the exploding alarm clock incident, now makes tempura batter as a divertissement. My dad, who was apprenticed as a foundryman and whose features were annealed in the glare of molten iron; he talks of the properties of soda water in the preparation of occult vegetables.

But it doesn't surprise me, because the kitchen is just another workshop, and cooking means using tools and materials to create something whole. Productivity, it seems, is something we crave. Fabulous kitchen knives are exquisite to behold and handle, and so are quality screwdrivers and spanners; so are needles and thimbles and bobbins to people who sew, and brushes and palettes to those who paint.

Reassembly has clear objectives – are there any bits left over, and does it work?

The problem is that reassembly, more than most other divertissements, reinforces a long-standing domestic divide. In the Andy Cap world, the woman remains in the house, cooking and sewing, while the man, in his sixth and seventh ages, slowly retreats to the shed or garage to fiddle with the lawnmower. Eventually, one or both of them dies, in the end unknown to one another.

This is why the reassembly of domestic appliances is a great fillip to modern, gender-fluid and mindful living. It begs to be done in the kitchen, in the spirit of communal and constructive living. There are plenty of things to be serviced and reassembled in there: toasters, coffee machines, juicers; even toasted sandwich makers, Sodastream machines and other things consigned to the under-the-sink cupboard of spent progressive fervour. They finally have a use.

Stimulate your partner's interest in reassembly by getting up in the night and taking some of your kitchen equipment apart. Do your bit to put the nuclear family back together. And the kettle.

'The world is
full of beautiful
electric guitars
nicely displayed
on stands in the
corners of sad
men's bedrooms...'

1984 TOKAI
ELECTRIC GUITAR

(147 PARTS)

JAPANESE COPY OF A FENDER STRATOCASTER

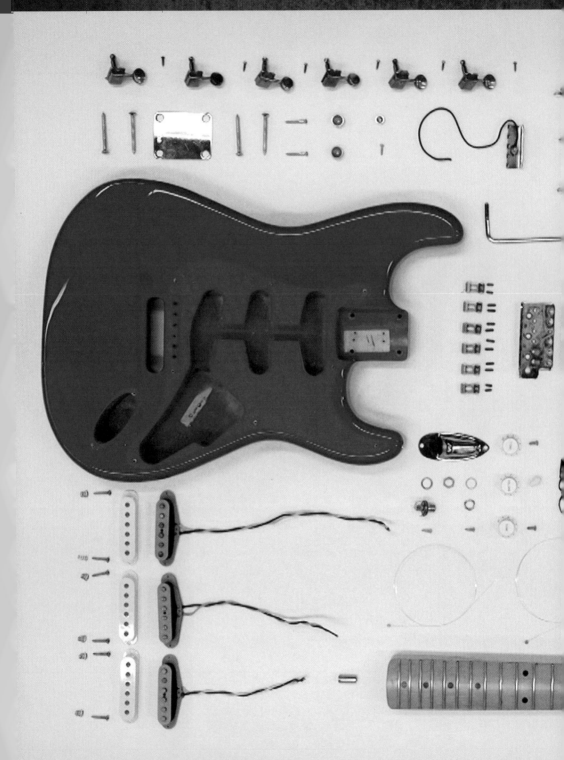

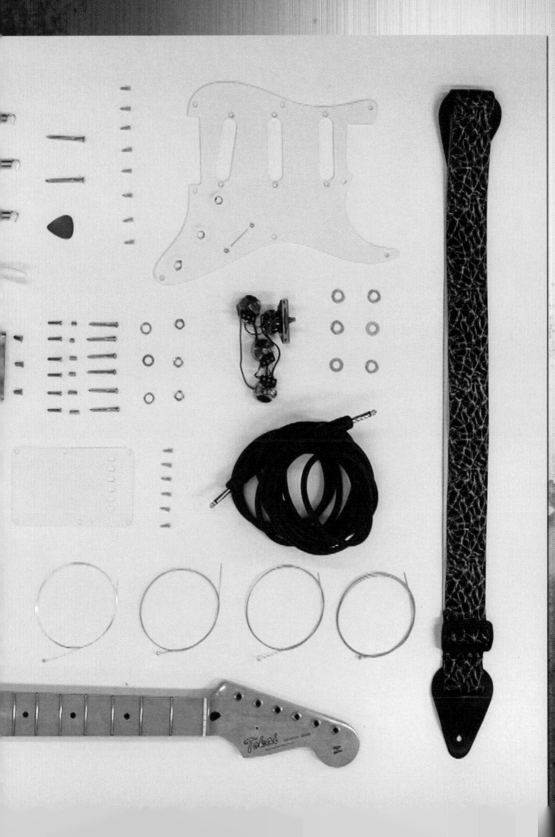

There came a glorious moment in the
reassembly of the electric guitar
when, after hours fussing over the trem
block, bridge plate, saddles, scratch plate,
whammy bar, tone controls and a multitude
of springs and screws, I threaded the first
string through the belly of the mute plank
and drew it up to the machine head.

I plucked it as I wound it on, obviously.
Some primaeval musical instinct, perhaps
related to the one that makes us twang a
ruler over the edge of a desk, forces us to
pick at anything apparently taut. There was
nothing at first, followed by a few miserable
metallic slaps, and then, suddenly, a thin but
seamless rising glissando. Music!

My reverie here was interrupted by the
sudden memory of the Rush album *2112*,
from 1976. In the story (this, regrettably,
being a 'concept album') the narrator, a man
living in a sterile future under the aegis of
the Priests of the Temples of Syrinx, finds
something odd in a cave behind a waterfall:

'It's got wires that vibrate and give music
What can this thing be that I found?'

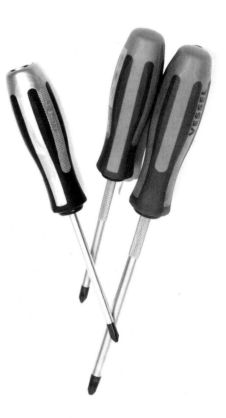

It's a guitar, strangely enough. He takes it to the priests, telling them that it's an 'ancient miracle' that can produce something 'as strong as life'. But they reject it, of course.

Progressive rock really is apocalypse-grade drivel, and is perhaps symptomatic of a serious problem with the electric guitar. To its credit, its invention, and the rock 'n' roll revolution that ensued, helped rid the world (America especially) of a plague of ukuleles, making it the Edward Jenner of musical instruments. On the other hand, it, along with the drum kit, is the most enthusiastically embraced yet least mastered of instruments, the vision of rock stardom perceived as something that can be achieved through mere ownership of equipment rather than by inspiration. Hence the world is full of beautiful electric guitars nicely displayed on stands in the corners of sad men's bedrooms that have only ever produced a few faltering bars of 'Stairway to Heaven'. A pull-down ladder to the attic would be more appropriate.

But back to reassembly. I was now communing with the ghost of Pythagoras,

More screwdrivers? Not quite. These are the Japanese Industrial Standard screwdrivers. See accompanying and deeply compelling article on page 138.

he whose experiments with strings, their lengths and tensions formed the basis of our understanding of pitch, interval and, ultimately, harmony. How apt.

The electric guitar, however, remains one of the less mystical of music-making devices. With the higher order of instruments – the violin, for example – we believe that, as with 'proper' food, it must be the work of a skilled old-world artisan to be credible. It must be created using techniques and understanding completely obscure to the rest of us, and will have a soul, and a unique voice. But the electric guitar is clearly just engineering science.

My Japanese copy of an original Fender is, especially, very obviously a manufactured good. Look at the parts. A tuning key operates a worm and pinion arrangement, exactly like the one between the electric motor and the driving wheels of the 00-gauge *Flying Scotsman* elsewhere in this book. The strings are indeed just lengths of high-grade wire. There are mundane potentiometers lurking below the volume and tone knobs. There are springs, levers, threaded adjusters and other things that you might easily find in a carburettor or a sewing machine. Even the cut-outs in the wooden body are obviously routed by a machine. This electric guitar employs some tiny grub screws, for Pete's sake. Grub screws are not romantic.

To speak at all, the electric guitar is wholly dependent on a basic bit of school physics that might have been better understood if it had been explained using one. When a magnet passes through a coil of wire, or vice-versa, a current is produced. This is the basis on which electric motors and generators work, the electromotive force.

The pick-ups of an electric guitar are simply magnets wound around many thousands of times with thin copper wire. When the string stretched just above this assembly vibrates, it sets up a disturbance in the magnet's field, inducing a tiny current in the coil. The strength of this current is related to the amplitude and frequency of the string's vibration, and is therefore properly analogue. These simple electric signals can then be passed to the stack of Marshalls for

If you can't actually play the guitar, it can still be used as the basis of an art installation.

amplification as *Ace of Spades*. Without this arrangement, an electric guitar would be just an ineffectual twangy stick.

Those signals can also be, and usually are, passed through things like effects pedals, so the sound of an electric guitar is not really 'true'. In a classical guitar, the instrument's tone is inherent in the shape and volume of its body, the materials it's made from and even the properties of the varnish used to finish it. In the pure electric guitar, electrical information coming down the lead is only the beginning of the adventure in sound.

(I am aware, of course, and in case you're gnashing your teeth furiously, that there are electric guitars that have pick-ups fitted to an acoustic body. But here we are concerned solely with the solid Tokai Fender Stratocaster copy)

To be honest, our guitar is the shape it is largely to make it more believable. There are no acoustic properties to the body; in fact, they are largely avoided in pure electric guitars, being a source of feedback generation. The eponymous Les Paul's first electric guitar had no real body at all, being based on a four-inch-square plank of pine. It was nicknamed 'the log', and audiences were apparently uneasy with it because they couldn't tell what it was.

So adding the vaguely guitar-shaped body is an exercise in making the thing more recognisably a guitar, nothing more. The 'wings' either side of the neck were ludicrously exaggerated during the era of glam rock, but were never really intended as 'wings' at all. They are cut-aways, there to enable the player's left hand to get at the highest frets. The body also provides a convenient mounting place for the controls and the jack-plug socket, but that's about it. The body is a genuine skeuomorph; a form retained because its familiarity establishes its function, rather than because it's actually needed.

Putting all this together is relatively straightforward and requires few tools – a handful of screwdrivers, an allen key or two and a slim spanner to tighten the securing nuts on the control pots. You will also have

This is a spring puller. It's for pulling springs. Tools are generally named unequivocally.

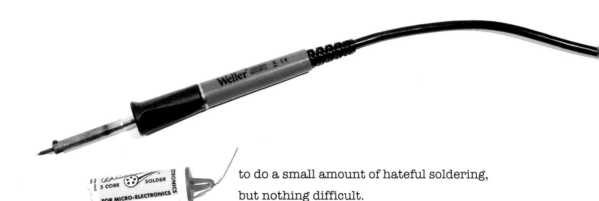

to do a small amount of hateful soldering, but nothing difficult.

The dreaded soldering iron and solder are the reason why swearing was invented.

That said, for the guitar to work to the satisfaction of an actual guitarist (I'm a pianist myself) it has to be set up very carefully. Each string passes through a saddle at the bridge end, which can be adjusted to raise or lower the string, minutely, in relation to the fretboard. The fretboard has a gentle transverse curve in it, making this trickier than it sounds.

The neck itself is also slightly arched longitudinally; by how much depends on the tension applied to the internal steel truss rod. This curve is then countered by the tension in the strings. But adjusting the truss rod can only be done before the neck is attached, so, as with the violin makers of Cremona, experience ultimately counts for everything. In music, technique is the liberation of the imagination. Technique, in turn, is liberated by proper set-up. This isn't as fiddly as a Boehm keyed flute or clarinet, but neither is it a piece of flat-packed furniture. It must be done properly.

What a joy, though, that the loveliest part of the process is necessarily saved until the end; attaching and tensioning the strings that elevate this mass of worthily engineered components to the status of instrument of music. This is where we came in. Science and technology had enabled art, and in return were granted the purity of purpose that art bestows.

It's a marvellous moment; a true transition, of the sort music should occasion within us, as components of humanity.

Note that this miniature pair of needle-nosed pliers are from the 'Precision range'. I don't have a pair from the vague range.

A constant supply of CO2 protects the musician from death by electrical fire.

JIS SCREWS IN FULL

The cross-head screw is a surprisingly complicated piece of manufacturing but an absolute godsend for the user, because your tool, your screwdriver, automatically centres itself in the screw, which is the best place for it.

The old-school slotted screw, of the type vigorously defended by 'proper' carpenters[1], allows the driver to wander about, slip, damage the surroundings and waste everybody's time.

So we're talking about a Phillips screw then. Or are we? First – and you might want to pour yourself a stiff one – some screw history.

The Phillips screw was invented not by a man called Phillips but by one John Thompson in the early 1930s. Having failed (like most inventors) to commercialise his idea, he sold it in 1935 to Henry F. Phillips, who in turn refined it and, crucially, persuaded General Motors to use it on automotive assembly lines. Later, it would become vital to wartime mass-production of aircraft, too.

---

[1] Mainly on the grounds of taste, it seems to me.

You can see why it would make sense, because in rapid-fire repetitive operations the self-centring screwdriver would save a fraction of a second every time it was used – it all adds up on production lines – and lend a reassuring certainty to the fidelity of screwed-together things.

But here's the bit most people don't know. There is an optimum tightness for screws in cars and aeroplanes, and achieving it requires some means of controlling the torque, the twisting force, applied to the screwdriver. On modern DIY power tools this is achieved with something like a rotating collar that causes an internal clutch to slip when the right torque is achieved. Back in the 1930s, though, that would have added cost and sophistication to the multitude of power tools being used in mass production.

So the Phillips screwhead is designed such that when the correct torque is reached, the screwdriver will automatically jump out of the slot, or 'cam out'. Every screw correctly tightened thanks to the carefully proportioned slot in the screw itself. The combination of that and self-centring meant the operator was required to exercise no more wit than that needed to point the screwdriver vaguely at the middle of the screw. Karl Marx would probably have something to say about that[2].

[2] Economic and philosophical manuscripts of 1844.

Meanwhile, Japan came up with its own universal cross-head screw, known simply as Japanese Industrial Standard, or JIS[3]. JIS screws are essentially different in that they are not designed to cam out. Instead, Japanese engineers concentrated on the size and, in some critical applications, even the metallurgical qualities of the screws themselves to prevent over-tightening. It takes a trained eye to tell the difference between a JIS and a Phillips screwdriver, but JIS screws are usually[4] identified by a tiny centre-punch mark or raised pip on the head.

Oddly, and perhaps portentously, a JIS screwdriver works perfectly in a Phillips screw, but a Phillips screwdriver will chew up a JIS screw.

That might not have mattered if history had worked out differently, but following Japan's post-war industrial boom, the world became full of Japanese electronic goods, motorcycles, cars, power tools and gardening equipment. It did not, however, become populated with JIS screwdrivers, which remain difficult to find in the US and Europe.

The net result is that generations of Western mechanics have cursed supposedly poor-quality Japanese fasteners when in

[3] Often referred to as 'Japanese Phillips'. But that's incorrect. So stop it.

[4] But not always, annoyingly.

The JIS screwdriver. In a workshop fire, it's the one you'd rush in to save.

fact the blame was more immediately to hand; that is, with the Phillips screwdriver.

My view is this. The camming-out attribute of the Phillips screw is actually its, er, undoing, as a stubborn Phillips screw will continuously reject its screwdriver when you try to remove it. It also doesn't help that Henry F. Phillips merely licensed his screw to other makers, and variations in the shape of the slot have crept in as a result, compounding the problem. I have never failed to remove a JIS screw if I'm using a JIS screwdriver.

Hunt them down and buy them. The realisation that the engine cases on your Honda lawnmower can, after all, be dismantled and reassembled without destroying the screwheads is like a new sun rising in your workshop.

(Newly ordained enthusiasts of screw and screwdriver history might also like to investigate Pozidriv, Frearson, Torx, Allen and Robertson.)

'As a very small child, I believed that words spoken into the phone stayed in there somewhere, rather in the way I believed that the Woodentops were actually inside the television, in an age before televisions became too thin for the idea to be credible.'

1957 GPO BAKELITE ROTARY DIAL TELEPHONE

(211 PARTS)

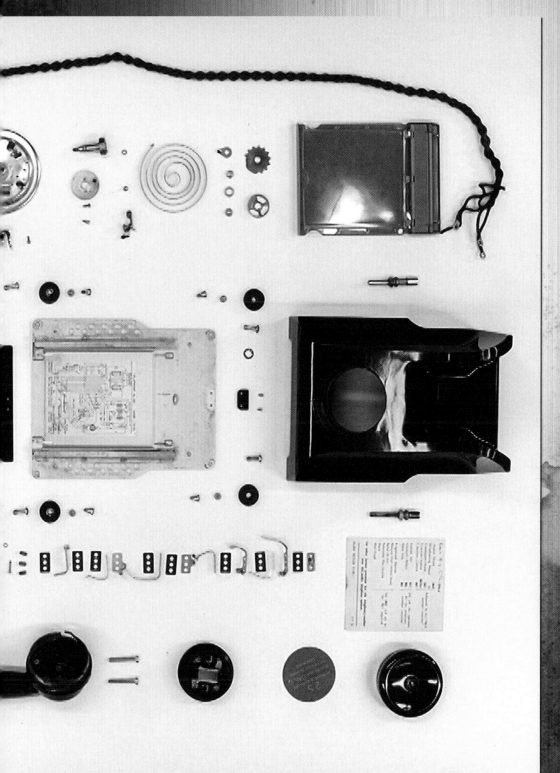

Old office equipment has always given me the creeps. Computer keyboards denuded of letters by furious fingers, seized mechanical adding machines, crackle-finish staplers and hole punches, superseded carbon copiers and dot-matrix printers, golfball typewriters, all pushed aside like an outdated combine harvester[1] to sit lifelessly exuding that peculiar smell of old kit, of imperceptibly decomposing paint and electrical insulation; useless testimony to the spent endeavours of people whose mouths are now stopped with dust[2].

I reserve particular abhorrence for the GPO model 332 telephone, the first British phone with an integrated and internal bell, introduced in 1937. It doesn't help that I dislike the shape of the 332, being much more of a fan of the later 706F MkII of 1966, with its rounded contours and cheerful colour schemes, or even the Trimphone. There's something oppressive about the (usually) black 332; something

[1] Phillip Larkin, 'High windows'.

[2] Omar Khayyam.

This handy small screwdriver incorporates a wire-stripping slot in its handle. Now if someone could just combine that with the handy pocket clip…

either Stalinist or funereal, two qualities I don't require of industrial design. This one lived somewhere in what the late Terry Wogan would have called 'the bowels of the BBC,' and who knows what passed through it?

For what is it? A conduit of rage, sharp talking, dealing, negotiation, panic and, yes, declarations of love. But those people are gone as well, their utterances lost to the vast entropy of telecommunicated passion. 'What will survive of us,' wrote the poet Larkin as he stared at the effigies on the Arundel tomb, 'is love.' In fact, all that is left of them is a slightly malodorous Bakelite monstrosity. Bakelite is such a miserable material, and a muted telephone is a horrible monument.

Maybe I feel this way because, as a very small child, I believed that words spoken into the phone stayed in there somewhere, rather in the way I believed that the Woodentops were actually inside the television, in an age before televisions became too thin for the idea to be credible. A part of me expects a disassembled old telephone to release the stale breath of the dead. Old telephones are haunted, by restless ghosts still seeking some sort of consummation.

So, to summarise, the old phone depresses me, until, that is, I look at its innards. It is an electrical device – powered by batteries in the exchange rather than the mains – and it includes a circuit board requiring enough soldering to induce self-harm. Much of that circuitry, however, is concerned with the integrated and internal bell, or rather bells, because the distinctive *ring ring* [pause] *ring ring* of a British telephone, as opposed to the blunt *riiiiiiiiiing* [pause] *riiiiiiing* of an American one, is as soft and melodious as it is because those two bells are tuned slightly differently[3].

Beyond that, the soul of the 332 is mechanical, and the parts are as polished

[3] Was the bell a bit of nominative determinism at work? Might history have turned out differently if the inventor had been called Alexander Graham-Kazoo?

and beautiful as the workings of a pocket watch. Consider the governor, in its dash pot, used to retard the return of the spring-loaded dial, a miniaturised rendering of a similar device used to control the speed of full-sized industrial machinery. Or the gleaming gravity switch plungers (Mk2), or the impulse contacts. The mausoleum of the 332 contains a hidden treasure of sparkling, if a bit fiddly, miniatures.

They do essentially one thing, which is to generate electrical pulses that travel down the line to alert the exchange or another telephone directly. How far you rotate the dial with your finger before releasing it dictates how many pulses are sent, the number of pulses being the number dialled. Achieving this reliably, consistently and at the right rate is a job for extreme mechanical propriety.

Then again, the mechanism is not burdened with any other duties. It allows you to dial another phone, and the circuitry makes your telephone trill enticingly when someone rings you. If you're in. If the phone rang in the olden days and no-one was there, did it actually ring?[4] There was no voicemail, so we can't be sure.

Less than a lifetime after Alexander Graham-Bell proclaimed that, one day, every town would have its own telephone, it had evolved into an instrument with an internal bell (The bells of earlier phones were often mounted on the wall, or even in another room entirely). About the same number of years after that it had become a fully portable device of such sophistication that people were using it to

The fibre pencil, or 'scratch pen', is perfect for polishing up small electrical contacts. I expect you want one now.

[4] Only if it made more noise than all the trees falling over.

write to each other. This is ironic in a way, as the telephone grew out of the telegraph, the meaning of which (without being too etymologically rigorous) is 'distant writing'.

But the telephone gave us a 'distant voice', which must have seemed a remarkable thing, because until it was invented everyone who had ever been heard had been within earshot. So no subsequent development in telephone technology – not even the iPhone battleships app – can ever equal the leap affected by the first one. Or two, strictly speaking[5].

It's worth contemplating all this as you stare, slightly baffled, at the pile of parts that turn out to be the innards of an old telephone. We may have lost sight of the value of the phone call itself, since being a phone is only one minor attribute of the things we carry around with us today. When using the phone was a thoroughly modern activity, there were educational films made

[5] History should record that Alexander Graham-Bell invented the first *pair* of telephones. His assistant Watson had the other one.

about how to do it, and how to talk to the operator, clearly; all that kind of stuff. It shouldn't surprise us that Alexander Graham-Bell came from a long line of elocutionists.

There was a time when the telephone was a votive object, representing personal advancement; when it was, according to one old-phone website, 'A status symbol, an item of pure luxury'. Let's not forget that the telephone once inspired a dedicated piece of furniture, the 'telephone table', so that your idol could be properly displayed, and you would have somewhere ruched and tasteless to sit while you genuflected towards it. But the telephone wasn't actually yours, because in the olden days all phones belonged (in the case of the UK) to the GPO, and the GPO would send a man, armed with exquisite screwdrivers (see accompanying article) to maintain and adjust it. It is, as I said, predominantly mechanical, and mechanical things need servicing.

So to reassemble the model 332 is to salute these men who kept us all talking. There are gears, cams, hairsprings, a clutch, detents –

familiar things in an unexpected role, there
to beget those pulses that once connected us.

And when it's complete, and the cryptic
workings are once again hidden away, you
realise that what has survived of it is just
the shape; a shape dripped into our sub-
conscious through old films, recreations of
wartime cabinet rooms, that sort of thing.
It has become part of the history of art
and design, while its pulses have been long
silenced by the new tonality.

Or have they? The fact is that the British
telephone network still supports pulse
dialling, even though it denies you the facility
to press the hash key for more options. With
the right plug fitted to the end of the lead, the
Model 332 (with integrated and internal bell)
will gift you the comforting purr through its
handset that says you have 'a line'.

Maybe, after all, an old telephone isn't
a tragic commemoration of those lost
exchanges. Maybe it's an admonishment
from history for the things we have left
unsaid. So I ring someone and tell her
I love her.

MAGIC SCREWDRIVERS

It's well known, I think, that the global ant biomass is greater than the human one. What is less well known is the total mass of lost small screws. So I've done a calculation.

If we assume that everyone currently alive in the world loses 25 grammes of small screws over a lifetime – and that's only a spoonful or so – then the globe is host to around 175,000 tonnes of missing screws.

Screws, by their nature, are standardised things, so ought to be easy to replace. Except that many have to be to specific lengths, or are in unusual small sizes, so unless you're one of those people who keeps spare screws in jam jars, you will probably never source a replacement. And if spare screws are being kept in jam jars, where have they come from? They have, at some point, been lost. So this screw-hoarding habit merely shifts the problem around and disguises it slightly.

That imagined mass of orphaned screws is a terrible indictment of society, because just as a kingdom can be lost for want of a horse, many a toy or domestic appliance has been consigned to the bin simply because a screw has fallen out and rendered it useless. Think of all the sunglasses in landfill, still perfectly able to block ultra-violet

My own screw-holding screwdriver. Until I had this, I couldn't hold on to any of my screws.

light but unable to stay on your face because one of those little screws that hold the arms on has dropped out, and no-one can get it back in even if they find it.

Not only do screws work loose, they can be snatched from us, cruelly, at the very moment of insertion, because they're too small to hold steady with the fingers, or they are to be located somewhere inaccessible. The answer is to hold the screw temporarily on the end of the screwdriver.

There are several ways of doing this. Some screwdrivers come with a soft plastic gripper on the end, although these may be fatter than the hole at the bottom of which the screw goes. Plasticine, Blu Tack, grease and paper glue can hold a screw on a driver, and so can spit if it's a really small one, but you might not want these things inside whatever it is you're reassembling. Especially if it's your watch.

You could use a magnetic screwdriver, or even magnetise one of your regular ones using a – wait for it – screwdriver magnetiser and demagnetiser. But magnetism can be your enemy (again, in your watch) and this won't work for brass screws, of which there are many in this telephone and in the toy train elsewhere in the book.

It may be time to consider the unequivocally named screw-holding screwdriver. This is a regular flat-bladed screwdriver in which the shaft has been sliced in two, lengthwise, and at a slight angle to the tip, forming two wedges. The two arms of the shaft are sprung apart lightly, and a sliding collar can be pushed down the shaft to bring them together.

So if the tip is inserted in a slotted screw, and the collar is pushed down, the two wedge-shaped ends slide over each other, effectively making the tip fatter, and the screw is held. It follows that the half-tips on the ends of each half-shaft must be hollow ground and not simply tapered like a normal screwdriver, otherwise this action would merely fire the screw off the end and we'd be back where we were at the beginning, adding to the universe's screw dark matter.

Screw-holding screwdrivers are strangely unknown to many amateur mechanics, but were essential tools for the GPO technicians who tended to the telephones of old. They were made by Quick Wedge of the US, the originators of the idea, and are often still found in the toolboxes of retired telecoms engineers, and therefore difficult to find. Quick Wedge are still in business.[1]

[1] Other makes of screw-holding screwdrivers are available. This book is brought to you in association with a BBC TV series.

A screw-holding screwdriver should not be used to fully tighten a screw, as the tip will be too weak. They are for locating screws. Finish the buggers off with a normal screwdriver.

As someone who has added more than his life's allotment to the lost screw continent, I can honestly say that the screw-holding screwdriver has improved my life and with it, the lot of humankind.

Incidentally, if that extrapolated figure of 175,000 tonnes of lost screws has alarmed you, please don't worry. The total mass of the planet hasn't changed, since the screws were wrought from the stuff of the Earth in the first place, have simply returned to it and remain a part of it. But that's precisely the problem. They were supposed to be part of the telephone.

999 – THE EXPLANATION

In that yawning moment of complete crisis, when the children are hanging from the window ledges of the blazing orphanage and you need to ring for the fire brigade, what you want is a nice memorable number that's easy to dial, obviously.

Millennials will doubtless be baffled by the idea of 'easy to dial', because on any vaguely modern telephone, all numbers are equally easy. But they weren't always equal. On a rotary dial 'phone, such as the GPO model 332[1], the number 1 is an easy outing for your index finger, but 0 will turn it into a bloodied stump.

Why, then, is the UK number for the emergency services 999? In the time taken for the ponderous dial to whirr its way back to rest, lives could be lost.

The first thing to know is that in the early automatic telephone exchanges, the electro-mechanical trickery needed a few pulses to come awake. That's why the area codes always began with a 0, which is actually 10 pulses. This wasn't an issue when the exchange was manned (or more usually, womaned) with people

---

[1] The first British telephone with an integrated and internal bell, in case I hadn't mentioned that.

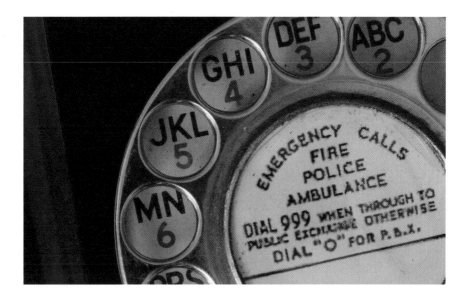

pulling out and pushing in jack plugs and saying 'You have a line, caller', and it wasn't a problem for local calls because they didn't go through an exchange and you could dial Wakefield One-Three[2] directly.

There is a second issue, however. Closely grouped overhead telephone lines could bang into one another during a storm, and this brief moment of contact produced a 'phantom' pulse. Using a three-digit number for emergencies reduced the chances of the wind calling for an ambulance, and making it a high number reduced it even further. The 10-pulse 0 was already taken for long distance, so the thinking was that 999 was difficult to dial accidentally, not like any long-distance dialling code, and statistically unlikely to be dialled by weather conditions. So 999 it is.

Thousands probably died as a result.

[2] Every really old person in Britain has a story about how their childhood telephone number was name-of-town two digits.

'A Monkey Bike
really is ludicrously
small, and sitting
on one is a bit
like using one of
those infant-school
lavatories as a
fully grown adult.
But it is,
unquestionably,
a genuine
motorcycle. '

HONDA Z50A
MONKEY BIKE

(303 PARTS)

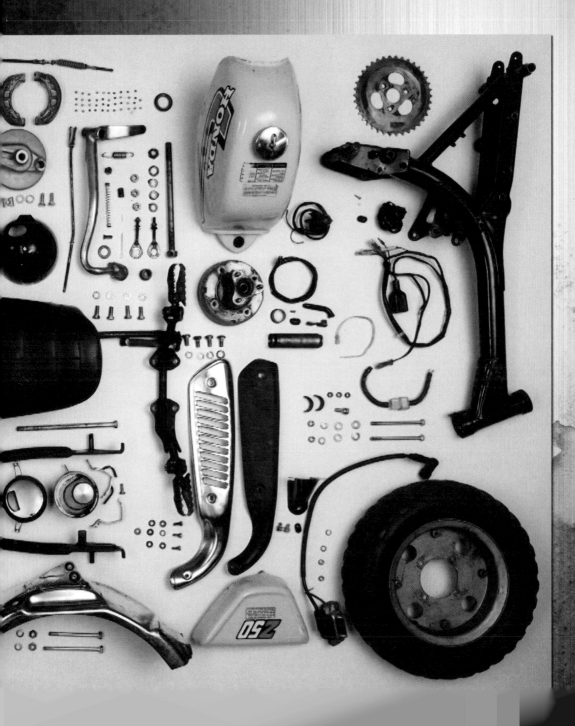

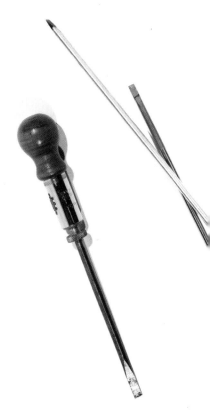

The beauty of the motorcycle, to my mind, is that it's the closest thing modern man has to the knight's charger of Arthurian legend.

The parallels are worth considering for a moment. You put on armour, and sally forth, often alone, astride a faithful companion and in search of enlightenment. The weather tyrannises you, and while you sit passively in a car, on a bike your mass and constantly changing bodily attitude are an essential part of the dynamics of movement. Rather as they are on a horse.

As a freshly licensed younger man, I would mount up and ride into the enveloping uncertainty of a winter's night to pursue women, believing they would swoon at my devotion; risking all upon the metal beast that they might grant me favour. A bit like Lancelot[1], though possibly only in my mind.

This view of motorcycling is entirely fanciful, but much more palatable than

Ratchet screwdriver (left) removes the requirement for you to take your hand off the screwdriver handle with each turn. Over a lifetime of workshop activity, this can represent several minutes of time saved.

[1] His very name suggests that he got more sex. I wonder if he was the inspiration for casting Roger Moore as James Bond.

the truth. Today, indeed, biking is a middle-aged, middle-class hobby, buoyed by hipster enthusiasm for specials, customs and other devices of style-conscious fair-weather fun. But for most of its existence the motorcycle was nothing more than a means of transport for people who couldn't afford cars. You will often hear motorcycling historians and the classic bike fraternity talk of 'ride to work bikes', the simple, unpretentious and largely forgotten machines that served us as a modest diesel hatchback does today. Not heroic, not legendary, not celebrated in verse or pre-Raphaelite art. Mostly just bloody miserable.

All of the above is a gross simplification of motorcycle history. There have always been glamorous bikes for hard-core addicts, and the ride-to-work bike is still very much with us. If you travel around places such as South-East Asia and rural India, where populations are dense, money less plentiful and the weather perhaps more conducive, you will find the small, simple motorcycle pre-eminent. In Vietnam, for example, there are so many two-wheelers on any road at any one time that a complex and bespoke code has developed to allow them all to interact safely and without hindrance. It's fascinating to watch, even more fascinating to take part in.

It should come as no surprise that the majority of bikes seen in these places are Japanese, copies of Japanese bikes, or at the very least heavily influenced by the breed. Japan's motorcycle industry achieved two great things during its meteoric rise in the 50s, 60s and 70s: they made the commodity motorcycle a much more dependable and efficient thing, and they, Honda especially, changed the image of motorcycling.

Back in biking's old world, we famously imagined that our planet-dominating industry was unassailable. We knew that the resurgent post-war Japanese industry was working on small bikes for the workers, but in our complacency imagined they would be content to do just that. We would build the prestige, high-value-added machinery that everyone coveted. Of course, Japan then had a crack at big machines, most famously the Honda CB 750 and the Kawasaki Z1, and showed that they could do it better. But,

The historic get-you-home slip pliers provided with small Hondas in the era of our Monkey Bike.

as with the great car makers, the Japanese bike factories earned their spurs[2] with humdrum stuff.

Honda earns special mention for its Cub[3] step-through utility bike, Soichiro Honda's riposte to the messy, noisy and unreliable two-stroke scooters and small bikes on sale in Europe, which he observed in the 50s on what would now be called a 'fact-finding mission'[4]. The Cub, in all its derivatives, has become the best-selling machine in the world, outnumbering the three best-selling cars in history combined.

The Cub was sophisticated, four-stroke, easy to use, economical and even, to some extent, weatherproof. More importantly, clever salesmanship by Honda in the US

[2] Note this second and subtle allusion to the knights errant analogy.

[3] These bikes are known variously as the Cub, Supercub, C50 and Dream, depending on where they were sold, trademarking being what it is. 'Cub' and 'Supercub' were already in use for Piper light aircraft in the USA, and 'Cub' was used by Triumph in the UK.

[4] Light spying, in industrial terms.

made it appealing to preppy kids, moms and other people who would previously have regarded a powered two-wheeler as the instrument of outlaws. So Honda became the family-friendly name in motorcycling, a bike for 'nice' people[5], and the Cub lent its reputation and essential componentry to the so-called Monkey Bike.

The Monkey Bike was also a massive hit in the States, because it was a bike for all the family to ride. It could be had with a semi-automatic three-speed box, it used the proven Cub engine, it was partly collapsible so it could be thrown in the back of a motorhome or station wagon, and it worked (to some extent) off-road. Or in 'the yard', if you were a typically land-endowed American.

Original 1960s Honda bike tool-kit spanner. Note the lovely logo forged into it.

It became known as the Monkey Bike simply because, according to legend, any adult astride its diminutive proportions and grasping the high bars assumes the attitude of a monkey riding a motorcycle. But then,

[5] 'You meet the nicest people on a Honda'.

A crimping tool, used to attach electrical connectors to the ends of wires without solder. For that reason alone it should be inducted into the tool hall of fame.

it was originally designed for children to ride, at Honda's own theme park[6]. That was the slightly ungainly (and now very rare) Z100[7] of 1964. By 1969 we had arrived at the second-generation Monkey, the Z50A, as reassembled here.

And here's my first reassembly tip. You could spend a lifetime reading about the decline and reinvention of the British and American motorcycle industries, about the rise of Japanese bikes, about how Soichiro Honda strapped small military surplus engines to bicycles to kick-start (push, actually) his empire, or how he took long sabbaticals to drink his own home-brew to the point of oblivion. Or you could just put a tatty old Z50A back together, because it will tell you everything you need to know.

[6] This was a shrewd move, as it meant the first bike that many future motorcyclists fell for was a Honda.

[7] Early Honda nomenclature is confusing. The Z100 had a 50cc engine, as did the C100 Cub. If it helps, the CB92 had a 125cc engine, but the C200 was a 90cc. Later, the number came to denote the engine size.

That the engine, for example, has a chain-driven overhead camshaft and a leak-proof crankcase, rather than the clanking (and rev-limiting) pushrods of the old British bikes, which wept fluids as if lamenting their own end. It displaces just 49cc, but the high-rev sophistication that Honda learned through racing made that just enough. Post-war, the resurgent Japanese automotive industry looked to efficiency and modernity to conquer a world increasingly concerned by fuel prices and pollution. Here is the evidence, in the engine of a child's motorcycle.

My travelling set of metric combination spanners in a handy carrying case. I usually travel without them.

That three-speed gearbox with its semi-automatic clutch is actually quite a complex assembly, but perfectly transparent in use. It relieves newcomers of the need to learn clutch control, which is one of the harder parts of learning to motorcycle. It's as easy as a twist 'n' go scooter but with the advantage of selectable ratios. Ideal for a pipsqueak off-road bike.

Elsewhere, there is evidence of foolproof Japanese production methods, now adopted the world over, in the way components go together the right way but not the wrong way. It's also surprising how few tools are needed for the job. The sort of toolkit found under the seat of a typical 60s or 70s Japanese motorcycle will probably do for 90 per cent of the work. Only the engine internals require anything specialised.

You will note, also, how one fastener on a pint-sized Honda will hold several things together, and how cables and wires are routed for neatness and to keep them away from the ill effects of the weather. Next to this sort of thing, a small British bike of the same era looks like an anachronism.

The most basic form of circuit tester. It's really just a light-bulb and two wires.

When it is finished – it should be possible within a day – you will have before you something remarkable. A Monkey Bike really is ludicrously small, and sitting on one is a bit like using one of those infant-school lavatories as a fully grown adult. But it is, unquestionably, a genuine motorcycle.

Perhaps the Z50 is a material manifestation of something very important to us. A modern, 200 horsepower, four-cylinder superbike represents an extreme; the

most potent resolution of a once simple idea, that of attaching a powerplant to a bicycle. So the Z50A, just about the least a motorcycle can be while still qualifying as one, represents an elemental truth.

The superbike will take you somewhere very quickly, but the 50cc single cylinder of the Monkey Bike will take you to the same place eventually, and dependably at that. That's why small-capacity Japanese motorcycles have trounced armies and political dogmas as the true liberators of people. You could ride a Monkey Bike across continents. People have done that.

Small, economical, dependable and all-terrain. You could end up like Sir Kay in the legends of King Arthur, wandering in the wilderness, your wits quite gone from you.

Wire strippers. The brass screw and lock-nut set the gap in the stripping jaws, for guaranteed repeatability of wire-stripping fidelity. Now you know.

You'd never guess from this photograph that the seat and the ride of a Monkey Bike were really comfortable.

THE GREAT BOLT VS SCREW DEBATE

What is the difference between a bolt and a screw? It's not entirely clear, and it should be made absolutely plain from the start that it matters less than a bent washer if all you're doing is mucking about in the garage with an old motorcycle. You probably won't have anyone to talk to anyway.

It's not likely to occasion any confusion. If you walk into a hardware shop and ask for 'some screws for my garden gate hinges', when actually you mean coach bolts, you'll probably be understood. If you say 'bolt' to a pedantic mechanic when really you mean machine screw, you'll probably be forgiven. It's not as if anyone is going to say 'vanilla slice' when they mean a threaded fastener of some sort, so there's no danger that we'll end up riding around on old motorcycles held together with bakery items. It's academic, but the muse of academia keeps our universities in business, and once in a while they come up with something like the transistor. So we'll keep going.

A lot of people reserve the word 'screw' for the tapered, coarse threaded excrescence used in wood, which in fact can share some attributes with a true bolt, if indeed we know what one of those is.

But we're not doing wood screws here. This is engineering, where everything threaded is of consistent diameter. Unless it's a metalworking self-tapping screw.

During broadcast of *The Reassembler*, one correspondent on social media told me emphatically that a screw is done up with a screwdriver and a bolt with a spanner, but – good news – it's not as simple as that. Screws can have hex heads and bolts can have screwheads. On motorcycles, you will find some screws, and bolts, with both a hex head and a slot for a screwdriver. So screw him, whoever he was.

A good beginner's definition is that a bolt is secured with a nut and a screw is screwed into another component. Please hold that thought for a moment.

To take this a bit further – and I know you'll want to – a true bolt features an unthreaded portion of its shank, which passes through plain holes in several components to bring them into alignment, and is secured by a nut, as on the garden gate hinges. This is what engineers would call a 'dowelling function'. So it would follow that screws are threaded for their entire length, and that is the difference. But it isn't.

It is if they're machine screws, which are threaded for their entire length. But what about the screws or bolts holding the side covers onto the Honda's engine? They have an unthreaded portion that locates in clearance holes in the cover, bolt-like, but they screw into tapped holes in the engine crankcase, which is another component. So if you go back a couple of paragraphs, you'll see that they're screws. They also have cap heads, so are tightened with neither a screwdriver nor a spanner, but with an allen key.

Here's another example. The rear mudguard of the bike is attached with hex head… things that are threaded for their entire (short) lengths, pass through clearance holes, and screw into the frame. Screws. Except what they screw into is a nut, welded in place. A bolt?

I'll leave you with this thorny conundrum. The front brake lever of the Z50 Monkey Bike pivots inside a double-sided bracket welded to the bars. A fastener, with a hex head and a plain portion to its shank, passes through the top half of the bracket and the lever itself, such that the lever pivots around the plain portion of the shank. But it threads into a tapped hole on the bottom half of the bracket, which is another component.

But then, because this is the front brake lever and you can't be too careful, it's secured with a nut.

I hope this has been useful.

THE REASSEMBLER
DISASSEMBLER APPENDIX

Most people know that the world of TV and film is awash with chauffeured limousines, Chablis Premier Cru on tap, caviar served in the navels of virgins, private jets, free cars, unchallenged admission to anywhere and all that sort of thing. The particular privilege I enjoy as presenter of the non-award-winning *Reassembler* series is a team of people on hand to dismantle any everyday object I care to mention at the drop of an M6 washer.

It would be easy to become lost in showbiz and forget that life isn't like this for most people. If you're going to have a crack at reassembling something, you'll probably have to take it apart first. You could accomplish this very quickly in a whirl of flashing spanners and screwdrivers, but where would that leave you? With a heap of meaningless bits, their relationship to one another lost to impetuousness; a tragic metaphor, perhaps, for the state of your own mind.

Better to do it systematically, so here are a few tips.

Firstly – and this will seem like a platitude – use the right tools and make sure they fit. Avoid Swiss Army things, cumbersome multi-bit screwdrivers, adjustable spanners, allen keys strung together on rings and all other instruments of the false prophets. Tools with a singular purpose always work better and are much nicer to use. Buy them separately, and relish their acquisition.

Now establish a 'geography of parts'. If disassembling the toy train, for example, decide that it's heading, say, from left to right on your workbench, and lay the bits out in that orientation as they come off. You should also lay bits out in the reverse order to the one in which they departed the whole, and in the correct spatial relationship to the other bits. That way you end up with a physical version of an exploded diagram. It'll look good, too; a sort of knolling.

It's not always exactly clear how things come apart. A case in point is old electrical goods, such as cassette recorders. There will be lots of small screws in the back, some holding the back on, but some holding internal assemblies in place. To establish which are which, loosen groups of screws by just a turn or so, and see what moves. If the back is starting to come loose, good. If it's still solidly in place but something is beginning to rattle around inside, bad. Do those screws back up, sharpish, and try a different group. Eventually you'll arrive at which screws need to come out altogether.

Screws (and bolts) often look identical from the head end, but can be different inside the darkness. Note, as you remove them, if any are longer or shorter than the rest. This is almost always true on engine cases of motorcycles and garden equipment. The geography of parts will help you here, as will mantras such as 'short at the front, long at the back', which also applies to German footballers' haircuts of the 1980s.

In reality, very few things have to be forced. In proper engineering, the size of fasteners is dictated by the magnitude of the job they're doing. So a huge bolt is doing a bigger job than a small one, but will be undone with a bigger, and longer, spanner. So the effort should be similar.

If things are stuck, it could be corrosion. But it could be because there's a lock nut somewhere. Or in the example of something like the left-hand pedal of a bicycle, you could be turning it the wrong way. Righty tighty lefty loosey doesn't apply to a left-hand thread, which is what you'll find on the pedal, to prevent it unscrewing in normal use. Righty will now be loosey.

If a part is secured by multiple screws, and the last one won't come out, the solution is sometimes to put the others back in and tighten them up, because it reduces the load on the remaining one. Screws stuck in place by corrosion can sometimes be loosened by first tightening them minutely, because it compresses the rust and breaks it up.

If something like a cover or casing won't come off, it could be because there is one more fastener lurking unnoticed in a dark corner. Does it move a bit, except at one end? Once you're absolutely sure there are no screws or clips working for the resistance, you can consider giving it a light smack.

What you don't need here is the claw hammer from your carpentry set. An old adage says 'When all you have is a hammer, everything becomes a nail', and things such as the die-cast aluminium alloy of the Honda Monkey Bike do not respond as a nail would. Although we think of metals as being one of the immutables of life, they are not. They break. Use a nylon-headed or wooden mallet if you can; if not, always put something like a piece of wood between the hammer and the hammered. There are whole motorcycles rotting in sheds because someone belted a stuck carburettor float bowl and then snapped off one of the slender pillars that support the delicate mechanism within. Remember what I said in the intro: machines are a bit like pets, and you are morally obliged to be kind to them. It's always worth asking yourself: would I do this to a puppy?

The hammer was probably humankind's first recognisable tool. The latest is the greatest thing bequeathed to the modern dismantler and remantler of things; the smartphone, because it has an instant camera in it. In the past, we had to make impressionist sketches or wait for real photographs to be developed by Boots the Chemist. Now you can keep a pictorial record of every stage of disassembly in your pocket, and look at it for succour in lonely moments.

Finally, remember that any fool can take something apart, but that it takes skill to reassemble. Dresden Cathedral was destroyed in one evening; it took 70 years to put it back together.

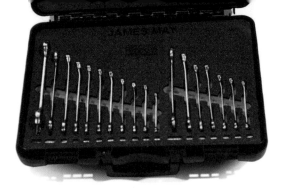

ACKNOWLEDGEMENTS

Cassian Harrison for commissioning the series for BBC4.

Plum Pictures for making it.

Brian Radman of the British Lawnmower Museum,
for supplying and dismantling the Suffolk Punch.

Gavin Payne of The Old Telephone Company, for supplying
and dismantling the GPO model 332.

David Silver and Dean Hendricks of the David Silver Honda
Collection, for supplying and dismantling the Monkey Bike.

Michael Stratton of Kenmix Engineering, for supplying and
dismantling the Kenwood Chef.

Mike Wood of Dansette Products Ltd, for supplying and
dismantling the Dansette Bermuda.

Malcolm Hine of MPH Guitars, for supplying and dismantling
the Tokai.

Rupert Lancaster of Hodder for not mentioning the awkward
missing book for the last three years; Cameron Myers,
Maddy Price and all at Hodder for making the book exist.

Bobby Birchall for the book design.

Hannah Wilcox at Plum Pictures and Lesley Hodgson
for additional picture research.

Ellis O'Brien for the tool pictures.

PICTURE CREDITS

All images in the book are courtesy of Plum Pictures except:

Images of individual tools by Ellis O'Brien ©Hodder & Stoughton 2017; Metal texture in background throughout ©Shutterstock; Page 31 ©Erik Strodl/fotoLibra; Page 39 ©Suffolk Iron Foundries of Stowmarket/1959; Page 43 British toolmaker Ray Willis ©Terence Spencer/The LIFE Images Collection/Getty Images; Page 57 ©Express Newspapers/Getty Images; Page 87 ©CW Images / Alamy Stock Photo; Page 89 Courtesy of Hornby Hobbies UK; Page 109 ©SSPL/Getty Images; Page 110 Image Courtesy of The Advertising Archives; Page 116 ©Shutterstock; Page 132-133 Tokai Brochure 1984/Special Thanks to Tokai Japan; Page 137©Fin Costello/Redferns/Getty images; Page 151 Courtesy of BT Heritage & Archives; Page 161 ©Lawrence Hill / Alamy Stock Photo; Page 181 ©Shutterstock.

ALSO BY JAMES MAY

Car Fever, How to Land an A330 Airbus,
James May's Magnificent Machines,
and *James May's Man Lab.*

First published in Great Britain in 2017 by Hodder & Stoughton
An Hachette UK company

1

Copyright © 2017 Plum Pictures Limited and James May

By arrangement with the BBC
The BBC logo is a registered trademark of the
British Broadcasting Corporation and is used under license.
BBC logo © 1996

A CIP catalogue record for this title is available from the British Library

Hardback ISBN 978 1 473 65691 8
Trade Paperback ISBN 978 1 473 65692 5
Ebook ISBN 978 1 473 65693 2

Typeset in American Typewritter and Avenir Lt
Designed by Bobby Birchall, Bobby&Co, London.

Colour origination by Born Group
Printed and bound in Italy by Graphicom S.r.l.

Hodder & Stoughton policy is to use papers that are natural, renewable
and recyclable products and made from wood grown in sustainable forests.
The logging and manufacturing processes are expected to conform to the
environmental regulations of the country of origin.

Hodder & Stoughton Ltd
Carmelite House
50 Victoria Embankment
London EC4Y 0DZ

*This book is not an instruction manual. We do not recommend that you
reassemble the projects yourself and if you do so then this is at your own risk.*

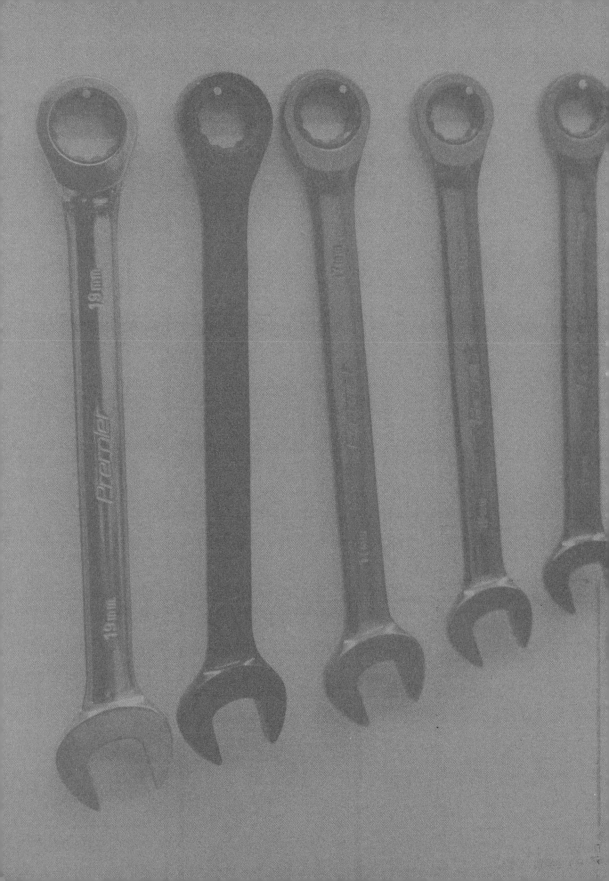